WORLDS Of DESIRE
THE CHICAGO SERIES ON SEXUALITY, GENDER, AND CULTURE
Edited by Gilbert Herdt

Also in this series
American Gay
Stephen O. Murray

Sexual Orientation
and the Military

EDITED BY

Gregory M. Herek
Jared B. Jobe
Ralph M. Carney

THE UNIVERSITY OF CHICAGO PRESS
CHICAGO AND LONDON

GREGORY M. HEREK is a social psychologist at the University of California at Davis. JARED B. JOBE is chief of the Adult Psychological Development Cluster at the National Institute on Aging of the National Institutes of Health. RALPH M. CARNEY is a social psychologist at the Defense Personnel Security Research Center.

The University of Chicago Press, Chicago 60637
The University of Chicago Press, Ltd., London
© 1996 by The University of Chicago
All rights reserved. Published 1996
Printed in the United States of America
05 04 03 02 01 00 99 98 97 96   1 2 3 4 5
ISBN: 0-226-40047-6 (cloth)
        0-226-40048-4 (paper)

The Library of Congress Cataloging-in-Publication Dara

Out in force : sexual orientation and the military / edited by Gregory
  M. Herek, Jared B. Jobe, Ralph M. Carney.
      p.   cm. — (Worlds of desire)
    Includes bibliographical references and index.
    ISBN 0-226-40047-6 (cloth : alk. paper). — ISBN 0-226-40048-4
  (pbk. : alk. paper)
    1. United States—Armed Forces—Gays. 2. Heterosexuals—United
  States—Attitudes. 3. Sexual orientation—United States. I. Herek,
  Gregory M. II. Jobe, Jared B. III. Carney, Ralph M. IV. Series.
  UB418.G38093 1996
  355'.008'664—dc20   96-20110
                                                            CIP

♾ The paper used in this publication meets the minimum requirements of the American National Standard for Information Sciences—Permanence of Paper for Printed Library Materials, ANSI Z39.48-1984.

To Jack and the boys—G.M.H.

*To my wife, Lisa; daughter, Jocelyn; son, James; sister, Karen;
and in memory of my parents, James and Jeanne—J.B.J.*

*To Ted Sarbin, who keeps psychology lively by
applying critical thinking to contemporary issues—R.M.C.*

Contents

Preface

The current volume was conceived after a workshop organized by Jared Jobe, with assistance from Ralph Carney and Gregory Herek, was held in conjunction with the 1994 national convention of the American Psychological Association in Los Angeles. The workshop papers were supplemented by papers prepared by several individuals affiliated with RAND at the time that institution's 1993 report to the Department of Defense was prepared. In addition, some new chapters were invited to increase the volume's comprehensiveness.

The book is divided into four parts. Part 1 provides a general overview of background information relevant to the policy. Chapter 2 describes the current state of scientific research on sexual orientation and behavior. Chapter 3 discusses the policy's legal basis and reviews the military and civilian court cases on homosexual military service and rights that relate to the current policy.

The chapters in part 2 provide comparative data relevant to understanding how gay men and lesbians might be successfully integrated into the military. The first two chapters review the military's experiences with integrating women and racial minorities, especially African Americans. The remaining two chapters describe the experiences of foreign militaries and domestic paramilitary organizations in confronting the issue of sexual orientation in their personnel policies and practices.

Part 3 considers social psychological issues relevant to the military policy. Chapter 8 reviews the extensive scientific literature on factors that affect social and task cohesion in the military and in comparable groups. Chapter 9 focuses on stereotypes concerning gay men and lesbians—how such stereotypes have formed and how they might be changed. Chapter 10 considers the "Don't Tell" component of the current policy. It reviews scientific data concerning the reasons that stigmatized individuals disclose information about their stigma to others, and describes the positive and negative consequences of such disclosure. Chapter 11 reviews the literature relevant to understanding personal privacy needs, including the ways that such needs are modified in response to changing circumstances.

Part 4 describes legal and implementation issues. Chapter 12 describes important issues of confidentiality, privilege, and breaches of confidentiality with military therapists, chaplains, and healthcare providers. Chapter 13 describes how a change in policy on gay men and lesbians—one that would allow open homosexuals to serve in the military—could be effectively implemented at the organizational level. Finally, chapter 14 reviews some of the political obstacles faced by President Clinton in attempting to overturn the previous military policy.

Although a variety of themes emerge from the different chapters, two of them warrant highlighting here. First, although the policy is ostensibly about gay people and homosexuality, it is actually about heterosexuality and the attitudes of heterosexuals. Defenders of the policy have generally conceded—often grudgingly—that many lesbians and gay men have served their country admirably in the military. The issue, they agree, is how heterosexual military personnel feel about serving with lesbians and gay men. The policy's proponents have characterized heterosexuals' negative attitudes toward lesbians and gay men as natural, normal, unchangeable, and even desirable. Opponents of the ban, in contrast, have argued that such attitudes are a form of prejudice or bigotry that has no more of a natural basis than do prejudices based on religion, race, or ethnicity—and is no more desirable than these other forms of prejudice.

A second theme is that the policy is not so much about individuals as it is about an institution—the U.S. military—and how that institution will adapt to changes in the larger society. As past experiences with racial and gender integration demonstrate, changes in long-standing prejudices and practices are not accomplished quickly or easily. Resistance to change from within and without will be intense and long-lived. Yet past experiences with other types of integration also indicate that the military is capable of meeting this challenge if it decides to do so.

The present volume is premised on the assumption that the social and behavioral sciences can make a valuable contribution to the policy debate by sorting opinion from fact, and myth from reality. It is the hope of the editors and contributors that the insights provided here will help to guide the military, lawmakers, the courts, and the president as they continue to grapple with this question of institutional and societal change.

The editors wish to acknowledge the invaluable assistance of Mary Ellen Chaney in preparing the final manuscript for publication. We also thank Doug Mitchell and Gil Herdt for all of their support and encouragement throughout the publication process, as well as two anonymous reviewers whose comments and suggestions greatly assisted us in completing the manuscript.

Gregory M. Herek
Jared B. Jobe

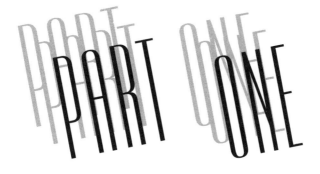

PART ONE

*An Orientation to
the Issue*

Social Science, Sexual Orientation, and Military Personnel Policy

Gregory M. Herek

Is the U.S. military capable of integrating openly gay personnel into its ranks while maintaining its ability to accomplish its mission?

In the courts, the Congress, and the mass media, proponents of the military's long-standing ban on homosexual personnel have answered this question in the negative. They have maintained that homosexuality and military service are incompatible. Until relatively recently, they based their arguments on the presumed failings of gay men and lesbians. Supporters of the ban maintained, for example, that gay men and lesbians are incapable of military service because they are not competent to perform their duties and because they pose security risks.

These arguments, however, have now largely fallen by the wayside. High-ranking military officers have stated publicly that gay male and lesbian personnel are generally competent at their jobs. In 1992, General Colin Powell, then chairman of the Joint Chiefs of Staff, said in congressional testimony that the reason for keeping lesbians and gay men out of the military "is not an argument of performance on the part of homosexuals who might be in uniform, and it is not saying they are not good enough" (House Budget Committee Hearing, 1992, p. 112; for empirical data supporting this point, see McDaniel, 1989). He further characterized individuals "who favor a homosexual lifestyle" as "proud, brave, loyal, good Americans" (House Budget Committee Hearing, 1992, p. 112). Similarly, the argument that lesbians and gay men pose a security risk has been abandoned (General Accounting Office, 1992a; Mills, 1995; Moskos, 1992; see also Herek, 1990).

⎾Instead of barring gay people from service on the basis of competence, more recent justifications have focused on heterosexuals' anticipated reactions to serving with openly gay and lesbian personnel. Concerns have been voiced that unit cohesion would decline, with a consequent reduction in unit performance; that heterosexuals' fears about privacy and bodily modesty would prevent them from sharing barracks, latrines, or foxholes with homosexual personnel; and that heterosexuals' antipathy toward homosexuals would lead to violence and breakdowns in the command structure. Opponents of the current policy have argued that these justifications simply reflect a willingness to tolerate the unfounded prejudices of the majority against a minority group. ⏋

Which side is right? Do arguments in support of the ban reflect valid concerns or are they rationalizations for personal and institutional prejudice? The present volume addresses this question, using theory and empirical data from the social and behavioral sciences. Scholars from the fields of psychology, sociology, law, and political science use the accumulated knowledge of their disciplines to provide a context for the current controversy and to consider the merits of arguments for and against the military's policies concerning sexual orientation.

Before undertaking this task, it is useful to place the debate in historical perspective. Since the birth of the republic, government decisions have been made about who shall be permitted or required to serve in the military, and under what conditions. These decisions have frequently reflected societal attitudes toward members of particular groups.

Early in the Revolutionary War, for example, black Americans were barred from service in the Continental Army. Similarly, Negroes were kept out of military service early in the Civil War, despite the eagerness of many northern blacks to volunteer (Stillman, 1968). Both policies were later reversed—when, respectively, the British began offering freedom to black slaves who would join their side, and the Union Army faced a serious shortage of troops (Binkin et al., 1982; Foner, 1974; Stillman, 1968). Even when they were allowed to serve, black soldiers were treated differently from their white counterparts. They were segregated from white troops and, when not in battle, often were assigned to menial occupations in peripheral units.

This pattern of racial exclusion and segregation continued through the first half of the twentieth century. Despite evidence that racially integrated combat units were effective in World War II (Stouffer et al., 1949), segregation remained official government policy until President Harry Truman's historic Executive Order 9981. Issued a few months before the

1948 election, Truman's order "declared to be the policy of the President that there shall be equality of treatment and opportunity for all persons in the armed services without regard to race, color, religion, or national origin" (Binkin et al., 1982, p. 26).

The armed forces subsequently began to institute a policy of racial desegregation, which dramatically increased the proportion of black service members by the Vietnam War era (Binkin et al., 1982). Desegregation proceeded slowly, however, and met with resistance (MacGregor, 1981). By the late 1960s, racial tensions had resulted in violent confrontations between blacks and whites, significantly affecting morale (Foner, 1974; MacGregor, 1981). The various service branches responded by instituting programs designed to address racial inequities and reduce interracial conflict. Although problems remain, the military's effort to confront institutional and individual prejudice within its ranks is widely regarded as a success story (e.g., Holmes, 1995).

The military has also attempted to integrate women into its ranks and to eliminate past patterns of institutional and individual sexism, although such efforts are less advanced than are efforts at racial integration. Since Congress passed the Women's Armed Services Integration Act in 1948 (Binkin and Bach, 1977; Stiehm, 1989), the role played by women in the armed forces has expanded considerably as a result of political pressure, legislation, court rulings, and Department of Defense (DoD) initiatives. The trend slowed during the 1980s under the Reagan administration, and some policies were reversed (e.g., Army basic training was resegregated). During the Bush and Clinton administrations, however, the role of women in the military expanded, with many women serving in the Persian Gulf during the 1991 war. Harassment of women has continued in the armed forces, however, as dramatized by the scandal surrounding the 1991 Tailhook Convention. Nevertheless, the military has maintained and enlarged its commitment to increasing gender equality.

In contrast to the armed forces' increasingly nonrestrictive policies concerning race and gender, their opposition to admitting and retaining gay male and lesbian personnel intensified after World War II. Before the 1940s, sodomy (usually defined as male-male anal sex, and sometimes including male-male oral sex) had long been considered a criminal offense and campaigns had been conducted periodically to purge military units of persons suspected of engaging in homosexual acts (Bérubé, 1990; Chauncey, 1989; Murphy, 1988). The military's prohibitions focused on homosexual behaviors, however, not on gay identity (Bérubé, 1990).

As psychological screening became a part of the induction process in the 1940s, the military adopted American psychiatry's view of homosexual behavior as an indicator of psychopathology. Based on a medical rationale, the military's focus shifted from conduct to status—from homosexual acts to homosexual persons. In 1942, revised Army mobilization regulations included for the first time a paragraph defining both the homosexual and "normal" person and clarifying procedures for rejecting gay draftees.

⌈As with black Americans, however, homosexual Americans were allowed to serve when personnel shortages necessitated it. When expansion of the war effort required that all available personnel be utilized, screening procedures were loosened and many homosexual men and women were allowed to enlist and serve. This shift was temporary. As the need for recruits diminished near the war's end, antihomosexual policies were enforced with increasing vigilance, and many gay men and lesbians were discharged involuntarily, usually with considerable stigma that followed them into civilian life. Throughout the 1950s and 1960s, acknowledging a homosexual orientation barred an individual from military service (see Bérubé, 1990, for a comprehensive history of the military's response to homosexuality during the World War II era).⌉

By the 1970s, however, a movement was emerging in the United States that pressed for civil rights for gay men and lesbians. The military policy was one target of this movement, dramatized by the legal challenge to the policy mounted by Leonard Matlovich (Hippler, 1989). Similar challenges continued throughout the 1970s. Although largely unsuccessful, they pointed to the considerable discretion allowed to commanders in implementing existing policy, which resulted in variation in the rigor with which the policy was enforced.

In 1982, the military's policy concerning homosexuality was revised. According to a General Accounting Office (GAO) report, the revision was undertaken primarily for three reasons: (1) to establish uniform procedures concerning homosexuality across the service branches; (2) to clarify the specific actions for which a person could be separated; and (3) to define the extenuating circumstances under which persons found to have engaged in those actions might nevertheless be retained (GAO, 1992a).⌊The 1982 policy stated that "Homosexuality is incompatible with military service. The presence in the military environment of persons who engage in homosexual conduct or who, by their statements, demonstrate a propensity to engage in homosexual conduct, seriously impairs the accomplishment of the military mission." This impairment was presumed to occur because

> [t]he presence of such members adversely affects the ability of the Military Services to maintain discipline, good order, and morale; to foster mutual trust and confidence among service members; to ensure the integrity of the system of rank and command; to facilitate assignment and worldwide deployment of service members who frequently must live and work under close conditions affording minimal privacy; to recruit and retain members of the Military Services; to maintain the public acceptability of military service; and to prevent breaches of security. (DoD Directive 1332.14, 1982)

Between fiscal years 1980 and 1990, the military discharged 16,919 men and women under the separation category of homosexuality. White women were discharged at a disproportionately high rate: 20.2 percent of those discharged for homosexuality were white women, although they constituted only 6.4 percent of personnel. In addition, the Navy was disproportionately represented, accounting for 51 percent of all discharges for homosexuality even though it comprised only 27 percent of the active force during this period (GAO, 1992b). These figures do not include involuntary separations of lesbians and gay men processed under other categories (GAO, 1992a).

Also during the 1980s, reversing the military's policy became a priority for advocates of gay and lesbian civil rights. Several lesbian and gay male members of the armed services came out publicly and vigorously challenged their discharges through the legal system. In 1992, legislation to overturn the ban was introduced in Congress by Representative Patricia Schroeder (D-Colo.) and by Senator Howard Metzenbaum (D-Ohio). By that time, nationwide opposition to the DoD's policy appeared to be increasing. Many national organizations had officially condemned the policy and many colleges and universities had banned military recruiters and Reserve Officers' Training Corps (ROTC) programs from their campuses in protest of the policy (Kosova, 1990).

By the beginning of 1993, it appeared that the military's gay ban would soon be overturned. Nine days after his inauguration, President Bill Clinton announced his intention to fulfill a campaign pledge by ending the exclusion of lesbians and gay men from military service. He asked Secretary of Defense Les Aspin to prepare within six months a draft of a policy to end discrimination on the basis of sexual orientation, and he proposed to use the interim period to resolve "the real, practical problems that would be involved" in implementing such a policy ("Clinton acts," 1993, p. A1).

Clinton's proposal was greeted with intense opposition from the Joint Chiefs of Staff, members of Congress led by Senator Sam Nunn (D-GA), conservative Republicans, and religious fundamentalists. After negotiation with Nunn and other members of Congress, the president publicly directed his secretary of defense to draft a plan for reversing the ban. In the interim, he ordered an immediate halt to the practice of asking new recruits about their sexual orientation, and he suspended discharge proceedings based solely on sexual orientation.

Nunn held widely publicized Senate hearings that exposed the public to the views of active-duty and retired military commanders, service personnel, academics, lawyers, and others, most of them opposed to lifting the ban (e.g., Schmitt, 1993a, 1993b). Hearings that were less lengthy and well publicized—but perhaps more balanced in their representation of viewpoints—were held by the House Armed Services Committee, chaired by Representative Ronald Dellums (D-CA). During the same period, the secretary of defense commissioned two studies of the policy, one by a panel of military personnel and the other by the RAND Corporation.

The RAND report (National Defense Research Institute, 1993) ultimately concluded that the military could successfully reverse the ban, provided sufficient leadership were exercised. The secretary of defense and the president, however, rejected most of the report in favor of recommendations by the military panel. They announced a compromise with Nunn, which they labeled "Don't Ask, Don't Tell, Don't Pursue" (Friedman, 1993, p. A14). Under its terms, personnel would not be asked about their sexual orientation and would not be discharged simply for being gay; engaging in sexual conduct with a member of the same sex, however, would still constitute grounds for discharge (Friedman, 1993). The compromise plan was widely perceived as differing from the former policy in only minor respects (Schmitt, 1993c). It was denounced by many lesbian and gay activists as a mere reformulation of the existing policy and likely to be unconstitutional. Since its enactment, numerous charges have surfaced that the policy does not protect gay and lesbian service members from harassment or witch-hunts, and that it has actually increased the number of discharges for homosexuality (Rosin, 1994; Schmitt, 1994, 1995; Shenon, 1996).

In the fall of 1993, Congress voted to codify most aspects of the ban, an outcome that the president initially had tried to avoid through his compromise. The congressional action was widely regarded as a retrenchment to an even more repressive military policy. One gay activist characterized it as "an ignominious defeat" for Clinton's compromise (Stoddard, 1993).

[The new policy contained four principal innovations. First, it distinguished homosexual orientation—which it defined as an "abstract desire"—from a propensity to engage in sexual acts with someone of the same sex. Second, it eliminated questions about sexual orientation from enlistment procedures. Third, it permitted associational activity, such as going to a gay bar or reading gay publications—provided that the individual did not indicate a propensity for homosexual conduct. Finally, it provided some discretion to commanding officers, which apparently allowed for more or less zealous attempts to separate gay men and lesbians from military service.

While the president and Congress have debated the policy, the judiciary's long-standing deference to military judgment and support of the ban have begun to break down. Several courts have expressed skepticism about the ban's necessity, both in its earlier form and in the "Don't Ask, Don't Tell" version. Early in 1993, for example, a federal district judge ordered the reinstatement of Petty Officer Keith Meinhold to the Navy and ruled that the ban was unconstitutional (Ferris, 1993). More broadly, he permanently enjoined the military from discharging or denying enlistment to any gay man or lesbian "in the absence of sexual conduct which interferes with the military mission" (Reinhold, 1993, p. A9). His broader ruling was reversed by a higher court, but Meinhold—like several other gay people whose discharges have been ruled illegal—continued to serve openly. Subsequent to Meinhold's case, a district court ruled that an acknowledgment by an individual service member of her or his homosexuality amounted to constitutionally protected free speech, and that the current policy was based on the prejudices of heterosexuals (*Able v. Perry*, 1995). The recent divisions of opinion within the judiciary make the policy a likely candidate for eventual Supreme Court review. The outcome of such a review cannot now be predicted.

The Larger Context of the Military Policy Debate

Throughout 1993, arguments in Congress and the media focused on the specific question of whether gay men and lesbians should be allowed to serve in the military. Slightly below the surface, however, raged a more general argument about the status of gay people in the whole of society.

At roughly the same time that the military policy debate was occurring, for example, conservatives and religious fundamentalists in several states were attempting to abolish local legislation that protects gay men and lesbians from discrimination. A ballot initiative with this goal was passed in Colorado in 1992 but declared unconstitutional by the Colorado Supreme Court. As this book was going to press, the U.S. Supreme Court upheld that ruling by a 6–3 majority. A similar ballot initiative was defeated statewide in Oregon in 1992, and again in 1994; subsequently, several local jurisdictions in that state passed their own versions of the initiative, which were declared invalid under state law. Supporters of the Oregon and Colorado measures have worked to pass similar ballot initiatives in Maine and several other states, so far without success.

Meanwhile, a Virginia circuit court upheld a lower court's decision taking a two-year-old boy away from his biological, lesbian mother and awarding custody to the child's maternal grandmother. Citing a 1985 Virginia Supreme Court ruling that a parent's homosexuality is a legitimate reason for losing custody, the circuit court judge characterized the biological mother as an unfit parent on the basis of her "illegal and immoral" conduct (Ayres, 1993). The Virginia Supreme Court upheld that judgment. Around the same time, Congress codified a ban on immigration by persons infected with HIV, the human immunodeficiency virus (O'Rourke, 1993).

In this context, the question of military service was probably not the arena that most gay and lesbian activists would have chosen for a national debate. Although the military has been a progressive force for racial equality in the past two decades, it is generally a conservative institution, especially in the realm of sexual politics. Like no other secular organization, the military remains a male-dominated institution whose members generally subscribe to the American ideology of masculinity and its attendant attitudes concerning gender and sexuality. Even as official policy has changed, individual attitudes have been slow to follow, as indicated by the military's ongoing struggles with the problem of sexual harassment.

Another problem with staging a fight for gay rights in the military arena was that, until very recently, the judiciary has historically deferred to the DoD in matters of civil liberties for military personnel. Individual rights that are guaranteed to civilians—such as freedom of speech and expression—are not always assured to the men and women whose primary mission is to defend those rights from foreign threats.

A third reason the military would not have been the first choice of gay and lesbian activists as an arena for a civil liberties debate concerns the military's historic relationship to politics in general. Support for a strong military establishment has long been integrally linked with anticommunism as the cornerstone of conservative political ideology. This connection reached new heights of prominence in the 1980s as the Reagan administration—citing the menace posed by the Soviet Union, which it characterized as an evil empire—committed unprecedented levels of funding to the Pentagon. Many liberals, in contrast, developed a deep distrust for the military during the Vietnam era. In the 1980s, they opposed the Reagan administration's defense buildup, arguing that it would greatly increase the likelihood of nuclear war and create a disastrous budget deficit that would ruin the domestic economy.

Thus, at the time of the national debate concerning gay people in the military, political conservatives (both leaders and members of the general public) had a strong ideological affiliation with the military whereas liberals generally did not. This was perhaps best exemplified by the frequent criticisms leveled at Clinton (first as a candidate, then as commander in chief), who had opposed the Vietnam War and did not serve in the military. Coincidentally, most conservatives have also opposed gay rights. Indeed, with the end of the Cold War and the U.S. Supreme Court's signals that it intends to uphold abortion rights, many conservatives focused on opposition to gay rights as a primary cause.

Liberals, in contrast, have been more sympathetic to gay men and lesbians, at least since the 1970s. But many liberals (gay and heterosexual alike) experienced ambivalence about the military issue as a result of their support for employment equality for gay men and lesbians, on the one hand, and their general antimilitary stance on the other. As a result, antigay conservatives were far more comfortable than were most gay-supportive liberals in arguing about how to maintain the military's morale and fighting effectiveness.

Despite all these problems, the military's ban emerged as the leading gay rights issue in the 1992 presidential campaign and the subsequent Clinton presidency. In retrospect, many explanations can be offered for what has happened since then. Opponents of the ban may have underestimated the difficulties inherent in overturning it. They may have naively assumed that, because the ban technically could be overturned by executive order, such an order could be issued and implemented by the president despite congressional opposition. Indeed, had Clinton been elected in 1992 with a strong majority of the popular vote (rather than a mere plurality in a three-way race), or had he been a conservative or a war hero, he might have been able to rescind the ban over the DoD's and Congress's objections. Even without these advantages, Clinton might have prevailed if he had been willing to expend the energy and political capital to do so.

But Clinton ultimately failed to fulfill his campaign promise. Many have interpreted his executive order as merely restating previous policy. Congress has enacted the ban into law, making it impossible for a future president to rescind it, even if he or she has the political will and ability to do so. The fight over the ban has now shifted to the judicial system.

Regardless of the eventual outcome in the Supreme Court, the recent debate surrounding personnel policy has had potentially lasting consequences. Political opposition to a nondiscriminatory policy appears to have solidified, and the political price paid by Clinton in attempting to change the policy may have been so high as to deter future presidents from initiating similar reforms for years to come. Yet recognition is now widespread that the military ranks include gay men and lesbians, even though much of the public prefers that they keep their sexual orientation a secret. The exemplary service records of many gay men and women have become public record, and past assertions that they lacked the skills and motivation to serve in the military have now been largely discredited. Even if military policy does not change, the breakdown of such stereotypes may provide a basis for gay rights advocates to attack civilian employment discrimination. The debate surrounding gay people in the military is probably best regarded as one small skirmish in the ongoing struggle to determine American society's attitude toward its lesbian and gay citizens.

References

Able v. Perry, No. 94 CV 0974 (E.D.N.Y. 1995).

Ayres, B. S. (1993, September 8). Lesbian loses custody of son to her mother. *New York Times*, p. A17.

Bérubé, A. (1990). *Coming out under fire: The history of gay men and women in World War II*. New York: Free Press.

Binkin, M., & Bach, S. J. (1977). *Women and the military*. Washington, DC: Brookings Institution.

Binkin, M., Eitelberg, M. J., Schexnider, A. J., & Smith, M. M. (1982). *Blacks and the military.* Washington, DC: Brookings Institution.

Chauncey, G., Jr. (1989). Christian brotherhood or sexual perversion? Homosexual identities and the construction of sexual boundaries in the World War I era. In M. B. Duberman, M. Vicinus, & G. Chauncey, Jr. (Eds.), *Hidden from history: Reclaiming the gay and lesbian past* (pp. 294–317). New York: New American Library.

Clinton acts to end ban on gays. (1993, January 29). *San Francisco Examiner,* pp. A1, A16.

Ferris, S. (1993, January 29). Gay sailor wins landmark ruling. *San Francisco Examiner,* pp. A1, A16.

Foner, J. D. (1974). *Blacks and the military in American history.* New York: Praeger.

Friedman, T. L. (1993, July 20). Chiefs back Clinton on gay troop plan. *New York Times,* pp. A1, A14.

General Accounting Office. (1992a). *Defense force management: DoD's policy on homosexuality.* (GAO/NSIAD-92-98). Washington, DC: Author.

———. (1992b). *Defense force management: Statistics related to DoD's policy on homosexuality.* (GAO/NSIAD-92-98S). Washington, DC: Author.

Herek, G. M. (1990). Gay people and government security clearances: A social science perspective. *American Psychologist, 45,* 1035–1042.

Hippler, M. (1989). *Matlovich: The good soldier.* Boston: Alyson.

Holmes, S. A. (1995, April 5). Time and money producing racial harmony in military. *New York Times,* pp. A1, A12.

Hope, R. O. (1979). *Racial strife in the U.S. military.* New York: Praeger.

House Budget Committee Hearing. (1992, February 5). *Reuter transcript report.* Reuters News Service.

Kosova, W. (1990, February 19). ROTC ya later. *The New Republic,* 24.

MacGregor, M. J., Jr. (1981). *Integration of the armed forces, 1940–1965.* Washington, DC: United States Army Center of Military History.

McDaniel, M. A. (1989). *Preservice adjustment of homosexual and heterosexual military accessions: Implications for security clearance suitability.* Monterey, CA: Defense Personnel Security Research and Education Center.

Mills, K. I. (1995, March 25). Gays can now get security clearance. *San Francisco Examiner,* p. A3.

Moskos, C. (1992, March 30). Why banning homosexuals still makes sense. *Navy Times,* p. 27.

Murphy, L. R. (1988). *Perverts by official order: The campaign against homosexuals by the United States Navy.* New York: Haworth.

National Defense Research Institute. (1993). *Sexual orientation and U.S. military personnel policy: Options and assessment.* Santa Monica, CA: RAND.

O'Rourke, L. M. (1993, March 12). Clinton loses vote on visitors with HIV. *San Francisco Examiner,* p. A14.

Reinhold, R. (1993, January 29). Military's gay ban loses in U.S. court. *New York Times,* p. A9.

Rosin, H. (1994, April 17). For gays in the military, the ban plays on. *San Francisco Examiner*, p. A15.

Schmitt, E. (1993a, April 1). Calm analysis dominates panel hearings on gay ban. *New York Times*, pp. A1, A7.

———. (1993b, May 11). Gay shipmates? Senators listen as sailors talk. *New York Times*, pp. A1, A9.

———. (1993c, July 22). New gay policy emerges as a cousin of status quo. *New York Times*, p. A7.

———. (1994, May 9). Gay troops say the revised policy is often misused. *New York Times*, pp. A1, C10.

———. (1995, March 13). The new rules on gay soldiers: A year later, no clear results. *New York Times*, pp. A1, A9.

Shenon, P. (1996, February 27). Homosexuality still questioned by the military. *New York Times*, pp. A1, A7.

Stiehm, J. H. (1989). *Arms and the enlisted woman*. Philadelphia: Temple University Press.

Stillman, R. J., II. (1968). *Integration of the Negro in the U.S. armed forces*. New York: Praeger.

Stoddard, T. (1993). Nunn 2, Clinton 0. *New York Times*, p. A13.

Stouffer, S. A., Lumsdaine, A. A., Lumsdaine, M. H., Williams, R. M., Jr., Smith, M. B., Janis, I. L., Star, S. A., & Cottrell, L. S., Jr. (1949). *The American soldier* (Vols. 1–2). Princeton, NJ: Princeton University Press.

Sexual Orientation and Proscribed Sexual Behaviors

Janet Lever and David E. Kanouse

In discussions of a policy change allowing homosexuals to serve, some of the most strongly expressed concerns have been that it would not only increase the number of homosexuals in the military, but also implicitly condone sexual behaviors now proscribed under Department of Defense Directive 1332.14 and Article 125 of the Uniform Code of Military Justice (UCMJ). The purpose of this chapter is to examine what we know about the prevalence of homosexuality and the proscribed sexual behaviors. Specifically, we review the best available data to answer these questions:

- What is the prevalence of homosexual behavior in the general U.S. population and in the military?
- Are homosexual status (i.e., self-identified sexual orientation) and homosexual conduct (i.e., sexual behavior) the same?
- What is the prevalence of the proscribed sexual behaviors among male and female heterosexuals and homosexuals?

Scientific Literature on Sexual Behavior

The literature on sexual attitudes, knowledge, and behavior contains serious problems, most of them unlikely to be resolved in the near future, if ever.

Much of the material in this chapter was drawn from an earlier chapter prepared by Janet Lever and David E. Kanouse in *Sexual orientation and U.S. military personnel policy: Options and assessments* (1993), Santa Monica, CA: RAND.

Virtually all available data from the time of Dr. Alfred Kinsey's pioneering works (Kinsey, Pomeroy, and Martin, 1948; Kinsey et al., 1953) until the past few years are derived from nonprobability convenience samples that are not generalizable to the U.S. population as a whole.[1]

In the past few years, researchers have attempted to apply random probability sampling techniques to get more representative respondents, but these studies, too, have serious limitations. Limitations are a result of sampling error, nonresponse bias, and various sources of measurement error, including the respondent's skipping embarrassing questions (item nonresponse), distortion of answers to fit a socially desirable image or to deny incriminating behavior, misinterpreting the question, or failure of memory to provide the accurate response. It seems likely, for example, that biases underlie the consistent finding that men report more lifetime sexual partners than do women. Reviewing these findings from four countries, Smith (1992, p. 210) concludes that much of the discrepancy is due to men's overreporting and women's underreporting; he raises the question "whether reliable and accurate sexual behavior data can be collected."

To date there is no completely accurate study of the prevalence and incidence of private sexual behaviors. Nevertheless, the data do provide some useful information regarding the three questions posed above. Fortunately, for most of the issues we examine, the information is adequate to compare sexual identity with behavior, to establish a lower bound for the prevalence of particular behaviors, and to estimate the relative prevalence in different populations.

In light of the variable quality of the research, we concentrate on those studies that provide the most generalizable empirical evidence available on issues relevant to this debate. Consequently, we give more weight to studies based on probability sampling methods than to those based on convenience samples. Priority is given to recent studies, which are more readily generalizable to today's policy context, all else being equal. We also rely on studies based on convenience samples when they include large numbers of people in groups underrepresented in probability samples, when they include detailed questions on behaviors of interest, or, simply, when no other data exist. In interpreting these data, we give attention to the probable direction of the biases.

Prevalence of Homosexuality: General Population and the Military

In some important respects, the prevalence of homosexual behavior in the general population has no direct bearing on policy regarding military service by homosexuals. If homosexuality is incompatible with military

service, then it is incompatible regardless of how many people are excluded by the ban (see chapter 8). Once consideration is given to ending
the ban, however, the prevalence of homosexual behavior gains relevance
from a practical point of view: How many potential military personnel
are we discussing? Accordingly, we review what is known about this
question.

Under current military policy, DoD Directive 1332.14 states that homosexuality is incompatible with military service. A homosexual is defined as "a person, regardless of sex, who engages in, desires to engage
in, or intends to engage in homosexual acts." As used in the directive, a
homosexual act "means bodily contact, actively undertaken or passively
permitted, between members of the same sex for the purpose of satisfying
sexual desires."

Simply put, DoD Directive 1332.14 prohibits *any* sexual contact between same-gender partners; it is the partner, not the act, who is proscribed.
But in applying DoD Directive 1332.14, the military recognizes the distinction between a homosexual orientation that is persistent and a single incident of homosexual conduct that is atypical of the person's usual conduct.
For example, if during an investigation it is determined that a homosexual
act was either a one-time "experiment" or the result of intoxication, adverse action need not be taken. Also, although the DoD definition includes
those who desire and/or intend to engage in homosexual acts, in practice
homosexual feelings are unobservable and exceedingly difficult to recognize in the absence of behavior and/or acknowledgment.

All of the post-Kinsey studies of the prevalence of homosexuality are
affected to some degree by problems of underreporting. Homosexual
behavior, especially in males, is highly stigmatized, and even the most
credible assurance of anonymity may not persuade survey respondents
to acknowledge behavior that they are accustomed to keeping secret.
Consequently, stigmatized sexual behavior is probably more often underreported than overreported, and the magnitude of the underreporting is
unknown.[2] Although much has been learned about survey research methods for obtaining useful data about sexual behavior, there are still many
unanswered questions about the effectiveness of different approaches
(Catania et al., 1990; Committee on AIDS Research and the Behavioral,
Social, and Statistical Sciences, 1990).

Homosexual Behavior in the General Population
Given these constraints, there is no definitive study establishing the proportion of men or women in the general population who ever or mainly

have same-gender sex. Instead, the proportion of men and women willing to acknowledge homosexual activity varies from survey to survey, no doubt in part reflecting the wording of questions on this highly sensitive topic (Catania, 1995) and possibly according to the methods used to assure confidentiality or anonymity.

If DoD 1332.14 were interpreted literally, the broad question of how many people ever or occasionally engage in homosexual acts is relevant. *Taken as a whole,* survey data indicate that roughly 2 to 8 percent of adult American males acknowledge having engaged in sexual acts with another man during adulthood. The extent to which the actual percentage may be higher, because of underreporting, is not known. For many men, long periods of time may elapse between such experiences. Consequently, the percentage of men who report such acts during specified periods (e.g., during the last year) is typically smaller than the percentage who report any such contact as adults. A majority of the men who report homosexual contacts have also had sex with women (Rogers and Turner, 1991). Thus, the percentage of men who are *exclusively* or *predominantly* homosexual in their adult sexual behavior (those most likely to consider themselves to be homosexual) is much smaller than the percentage who ever have sex with other men. (We discuss this issue further under "Relationship Between Status and Conduct.")

Data on the prevalence of female homosexuality are even more sparse than data for males, and where data have been collected, they are often unreported. The data that exist, however, suggest a prevalence lower than for males: The estimates of those who ever engaged in sexual acts with another woman during adulthood range from 1 to 6 percent, with variations among age groups and for marital status.

For many years, virtually the only data came from Kinsey, Pomeroy, and Martin (1948, p. 651), the source for a widely cited figure of 10 percent. In fact, this figure referred to the estimated proportion of the fifty-three hundred men interviewed who were exclusively or predominantly homosexual—*for at least three years between the ages of sixteen and fifty-five.* They estimated the proportion who were exclusively homosexual throughout their lives to be much lower—4 percent.[3] Kinsey et al. (1953) are often cited to the effect that the prevalence is lower among females than among males. But such a conclusion requires comparable data for both genders, and unfortunately, Kinsey et al. did not report on female homosexual behavior using the same yardstick as was used for males.[4]

More recent data from probability samples suggest that Kinsey et al.'s (1948) 10 percent figure for males is too high. But recent studies, summarized in the table below, still do not converge on a single "correct" figure below that number. The prevalence estimates shown in the Table on page 20 are not directly comparable to Kinsey et al.'s 10 percent figure. Rather, the statistics refer to all those who report any same-gender sexual contact either in adulthood or during a specified time period—a number likely to be considerably higher than the percentage who report being exclusively or predominantly homosexual.

With a narrower interpretation of DoD 1332.14, the focus shifts from those who ever or occasionally engage in homosexual acts to those who routinely do so. Ideally, we would estimate the proportion of the population that is predominantly or exclusively homosexual. That proportion would no doubt be greater than the proportion that is exclusively homosexual.[5] Unfortunately, surveys based on probability samples have asked questions that provide data more directly comparable to Kinsey's 4 percent estimate of those who have been exclusively homosexual throughout their lifetime.

In the National Health and Social Life Survey (NHSLS) of a representative sample of 3,432 men and women aged eighteen to fifty-nine, 0.9 percent of men and 0.4 percent of women report having only same-sex partners since the age of eighteen (Laumann et al., 1994). Of the 3,321 men aged twenty to thirty-nine surveyed in the National Survey of Men (NSM-1), only 1 percent reported being exclusively homosexual in behavior in the prior ten years (Billy et al., 1993).[6] In their reanalysis of five probability studies (all presented in the table), Rogers and Turner (1991) note that only 0.7 percent of men report exclusively male-male sexual contacts during adult life. Results from these studies are strikingly similar.

The table clearly indicates the episodic or experimental nature of homosexual experiences for some people. The shorter the reference period investigated, the smaller the percentage of men and women who report any particular sexual behaviors, including same-gender contact. Besides the reference period, differences in samples and data collection techniques likely also contribute to the variation in prevalence estimates. Estimates of homosexual activity are highest in the Research Triangle Institute (RTI) study, which was conducted as a pilot test for a national seroprevalence study (Rogers and Turner, 1991). Its unusually high response rate (88 percent) may be a result of the cash incentive offered; in addition, it is possible that a higher proportion of homosexual men agreed to partici-

Estimates of Homosexual Behavior from U.S. Probability Studies

Study	Sample Characteristics	Prevalence of Any Same-Gender Sexual Contact	Methods of Data Collection	Response Rate
National Opinion Research Center (NORC) 1970 (Fay et al., 1989)	1,450 men aged 21 and older	Since age 20 Males: 6.7% Females: N/A	SAQ following face-to-face interview	N/A
		Last year Males: 1.6–2.0% Females: N/A		
General Social Survey (GSS)* 1989–1991	1,564 men and 1,963 women aged 18 and older	Since age 18 Males: 5.0% Females: 3.5%	SAQ following face-to-face interview	74%–78% (1988–1991)
	1,941 men and 2,163 women aged 18 and older	Last year Males: 2.2% Females: 0.7%		
Louis Harris & Associates, 1988 (Taylor, 1993)	739 men and 409 women aged 16 to 50	Last 5 years Males: 4.4% Females: 3.6%	SAQ following face-to-face interview; same-sex interviewer	67%
		Last year Males: 3.5% Females: 2.9%		
		Last month Males: 1.8% Females: 2.1%		
Research Triangle Institute (Rogers and Turner, 1991)	660 male residents of Dallas County, Texas, aged 21–54	Last 10 years Males: 8.1% Females: N/A	SAQ	88%
		Last year Males: 4.6% Females: N/A		
National Survey of Men (NSM-1) (Billy et al., 1993)	3,321 men aged 20–39	Last 10 years Males: 2.3% Females: N/A	Face-to-face interview; female interviewers	70%
National Health & Social Life Survey (NHSLS) (Laumann et al., 1994)	1,410 men and 1,749 women aged 18–59	Since age 18 Males: 4.9% Females: 4.1%	SAQ following face-to-face interview	79%
		Last 5 years Males: 4.1% Females: 2.2%		
		Last year Males: 2.7% Females: 1.3%		

Note: N/A = not available. SAQ = Self-administered questionnaire.

*Prevalence of male and female homosexuality calculated at RAND from General Social Surveys (Davis and Smith, 1991).

pate because of the survey's AIDS focus. In any case, its sample is composed only of residents of Dallas County, Texas. There is no reason to believe that the true prevalence for Dallas County mirrors that of the nation as a whole.

One would expect higher prevalence from the RTI study insofar as results from several studies listed in the table show strong consensus that male-male sexual activity is reported more frequently in urban than nonurban areas. By pooling the data collected from the General Social Survey (GSS) since 1988 with those from the NHSLS to increase sample size for the analysis, authors of the NHSLS report on the effect of urban/rural place of residence (Laumann et al., 1994). Men living in the central cities of the twelve largest metropolitan areas report rates of same-gender sexual contact since age eighteen of 16.4 percent, compared with rates of 1.5 percent in rural areas and 5.0 percent in the total sample. A reanalysis of some of the other probability surveys listed in the table also shows higher rates in cities (Rogers and Turner, 1991).

Estimates of homosexual activity are lowest in the National Survey of Men (NSM-1), but data collection proceeded differently from all other surveys presented in the table. Whereas the other surveys used a self-administered questionnaire for sensitive questions that was completed and delivered in a sealed envelope to the interviewer after a face-to-face interview, the NSM-1 was conducted only with face-to-face interviews. Moreover, in contrast to the use of interviewers of both genders, or ones matched by gender to the respondent, the NSM-1 used all female interviewers for all male respondents. These methodological variations may account for the low rate of reported homosexual behavior.

Finally, differences in prevalence estimates may be due to sampling and/or measurement error. First, no sample perfectly represents the population from which it is drawn, so statistics are often reported using confidence intervals that estimate the likely range of variation due to sampling error. Where confidence intervals are offered, there is much more overlap between study estimates.[7] Second, estimates may be affected by low response rates. Rates for the surveys shown in the table ranged from 67 percent to 88 percent; although these are considered acceptable rates for in-person household surveys, they still imply that between one and three of every ten persons refused to participate.

There is no evidence to show whether persons with homosexual experience differ from those without homosexual experience in their willingness to cooperate with survey researchers. As we discussed earlier, however, it is likely that many of those with homosexual experience who do participate in surveys do not acknowledge that experience; this underreporting is one component of measurement error. According to the president of Louis Harris and Associates, measurement error is a far bigger problem than sampling error when there is a socially desirable answer in surveys of both behavior and attitudes (Taylor, 1993).

The extent of measurement error is unknown. Reanalyzing 1970 data collected by the National Opinion Research Center, Fay et al. (1989) appropriately suggested that their estimates be viewed as "lower bounds on the prevalence of same-gender sex among men" (p. 243).[8] Other scientists concur that such estimates should be thought of as lower bounds of actual prevalence (Laumann et al., 1994; Rogers and Turner, 1991). Nevertheless, the new probability studies indicate that the current prevalence of male homosexual behavior and of exclusively homosexual behavior patterns are lower than Kinsey et al.'s (1948) widely cited estimates.

Homosexual Behavior among Military Personnel

Few studies have asked military personnel about their sexual activities, and none have published data on the incidence of homosexual behavior among those currently serving in the armed forces. The only study from which an inference can be made is based on three national probability samples that included data on previous military status. These data suggest that the prevalence of same-gender sexual behavior by men who have served in the military is at the high end of the range for the general population (Rogers and Turner, 1991). This behavior may or may not have occurred during their military service.[9]

Rogers and Turner (1991) report an analysis combining data from three probability samples of the U.S. population (combined $n = 2,449$ respondents) that examines the proportion of men aged twenty-one and older who reported adult same-gender sexual experience by various social and demographic characteristics, including military service. Among men with military service, 7.6 percent reported same-gender sexual contact, compared with 5.1 percent of other men. Military service was one of only four *adult status* variables that showed a reliable statistical relationship with reports of same-gender sex across the three surveys.[10]

Relationship between Status and Conduct

Under current military policy, there is a "rebuttable presumption" that homosexual status equals conduct: A soldier can be discharged either for being homosexual or for engaging in a homosexual act. In this section, we review studies of sexual behavior and/or identity to explore whether homosexual status and conduct are the same. If the two are not the same, then a policy of excluding solely on the basis of status would exclude some who do not engage in sexual acts with same-gender partners, while allowing others who do to serve. In this chapter, we do not address the policy problems that exclusion on the basis of status might pose, but merely the question of how many people might fit the broad DoD definition of homosexuals. Moreover, this section has implications for health-related concerns, because it is conduct rather than status that poses potential health risks.

A review of available studies leads us to conclude that whereas there is a strong correlation between status and conduct, they are not the same thing:

- A person who *does not* identify himself or herself as a homosexual may still engage in acts with someone of "the same sex for purposes of satisfying sexual desires" (in the language of the directive);
- A person who *does* identify himself or herself as a homosexual may refrain from engaging in homosexual acts.

Homosexual Behavior among Self-Identified Heterosexuals

Kinsey et al. (1948) used "homosexual" or "heterosexual" not as nouns characterizing people but as adjectives characterizing acts. In their landmark study, they created a seven-point scale—which came to be known as the "Kinsey scale"—to place individuals along a continuum ranging from exclusively heterosexual (0) to exclusively homosexual (6), according to his or her current or cumulative lifetime sexual experiences and sexual feelings. All intermediate points indicated personal histories with a mixture of homosexual and heterosexual acts and/or feelings. Kinsey et al. (1948, p. 650) found that most of those who ever engaged in homosexual acts had engaged in more heterosexual than homosexual acts. The studies based on probability samples presented in the previous section support the generalization that a majority of men who report male-male sexual contacts in adulthood also report female sexual partners in adulthood (Rogers and Turner, 1991, pp. 505, 509).

After analyzing the sex histories of one hundred fifty respondents who had both heterosexual and homosexual experience in adulthood, Blumstein and Schwartz (1976, p. 342) concluded there may be "little coherent relationship between the amount and mix of homosexual and heterosexual behavior in a person's biography and that person's choice to label himself or herself as bisexual, homosexual, or heterosexual."

The relationship between identity and behavior has not been well studied, because the available data sets have generally included measures of only behavior or identity, or have been based on very small and nonrepresentative samples. The NHSLS is the exception; in addition to questions on identity and behavior, there are two items that serve as an indicator of homosexual desire. Identity, behavior, and desire coincide in a core group of 2.4 percent of the men and 1.3 percent of the women, but the authors say that "there are also sizable groups who do not consider themselves to be either homosexual or bisexual but have had adult homosexual experiences or express some degree of desire" (Laumann et al., 1994, p. 301).

Another data set that contained independent measures of behavior and identity on a large national sample of 56,600 men supports the conclusion that conduct and status are not the same (Lever et al., 1992). Lever et al. reanalyzed a 1982 readers' survey that appeared in *Playboy*. Obviously, respondents are not representative of all U.S. men (nor even all *Playboy* readers), so the data set cannot be used to estimate prevalence of behaviors, but it can provide information on the relationship *between* various aspects of sexuality. Accordingly, Lever and colleagues examined the 6,982 men (or 12.5 percent) who reported adult sexual experiences with both men and women. Of these, 69 percent described themselves as "heterosexual," 29 percent as "bisexual," and 2 percent as "homosexual." Even after allowing for likely overrepresentation of men at the heterosexual end of the Kinsey et al. (1948) continuum, the result demonstrates that many men who have engaged in homosexual conduct do not consider themselves homosexual.

An epidemiologic study and a criminological study further illustrate the point that homosexual behavior does not occur only among people with homosexual identification. Researchers (Doll et al., 1992) from the Centers for Disease Control and Prevention studied 209 HIV-seropositive male blood donors who reported having had sex with both men and women since 1978. Because men who have had sex with men are asked to refrain from donating blood, one might expect this sampling method to overrepresent men who do not have a homosexual self-identification. Of these, 45 percent self-identified as homosexual, 30 percent as bisexual, and 25 percent as heterosexual.

Studies in criminology have found examples in prison of what social scientists term "situational homosexuality," that is, self-identified heterosexuals engaging in homosexual behavior only in situations that preclude sex with women. Wooden and Parker (1982) conducted what is considered to be the most thorough study of the phenomenon of male-male sexual activity in a prison context. Through in-depth interviews, the researchers learned that the sexual aggressors consider themselves "heterosexual"; their targets are men they assume to be homosexual or younger heterosexual men who are not able to protect themselves. Most of the sexual aggressors claim no homosexual experience before prison, and those released claim to resume a life of exclusively heterosexual relations. Of the two hundred men in Wooden and Parker's study who returned a questionnaire, 10 percent identified themselves as homosexual, 10 percent as bisexual, and the remaining 80 percent as heterosexual; more than half (55 percent) of the heterosexual group reported having engaged in homosexual activity in prison.[11] Although prison culture may create unusual incentives for situational homosexual contact, these behavioral patterns offer another example of divergence between identity and behavior.

Virginity and Celibacy among Self-Identified Homosexuals

Current military policy considers that a statement of homosexual orientation presumes homosexual behavior. Therefore, we examined various studies of whether people may have a sexual self-identification that incorporates attraction to others of the same sex without having acted on their homosexual feelings. We use as examples two studies with probability samples—one a national sample of male adolescents and one a single-city study of homosexual and bisexual men—as well as an epidemiologic study and three nonprobability surveys, one of homosexual men and two of homosexual women.

In 1988, the Urban Institute conducted a nationally representative survey of adolescent males, which included a self-administered questionnaire that contained sensitive items on sexual practices. Researchers at RAND used the machine-readable data file (Sonenstein et al., 1991) and adjusted by case weights to explore the item on sexual identity. Of the 1,095 males between ages seventeen to nineteen in the total sample, 86.4 percent answered "heterosexual," 5 percent answered "mostly heterosexual" or "bisexual," and 0.6 percent answered "mostly homosexual" or "100 percent homosexual" (8 percent answered "don't know" or left the item blank). Only 23 percent of those whose identity was not exclusively heterosexual had ever engaged in sexual acts with another male—that is, roughly three-quarters were "virgins" with regard to homosexual sex.

Very few studies of homosexual men are, like the Urban Institute study, based on a probability sample screened from a random sample of the general population. One study used a probability sample, but not from the general population. That study was conducted by RAND in 1989–1990 of three hundred homosexual and bisexual men over age eighteen who were concentrated in areas of Los Angeles County known to have significant numbers of homosexual men (Kanouse et al., 1991a). The sample included men who acknowledged having had sex with another man in the last ten years. Although this study overrepresents men living in homosexual neighborhoods relative to those living in other areas, the sample is in other respects likely to be much more representative of homosexual men than, say, a sample of men attending a sexually transmitted disease (STD) clinic or men who belong to a homosexual organization. In an anonymous telephone interview, homosexual and bisexual men were asked detailed questions about their sexual risk behaviors. About 13 percent of respondents reported having no sexual partner of either gender in the past twelve months.[12]

Another probability study of homosexual and bisexual men conducted in San Francisco shows a similarly high rate of sexual inactivity for a large minority of men (35 percent) when a short reference frame is used, in this case, the past thirty days (Stall et al., 1992). In a nonrepresentative sample of 584 homosexual and bisexual men recruited in "gay" establishments in Pittsburgh likely to overrepresent sexually active men, 8.2 percent reported that they had been celibate for the previous six months (Valdiserri et al., 1989).

Loulan (1988) distributed sex questionnaires at therapy workshops, lectures, and women's bookstores as well as through advertisements in women's and homosexual newspapers throughout the U.S.; we assume that her sample overrepresents women who are openly homosexual and part of the visible homosexual community. Self-reported histories of the 1,566 homosexual women who responded showed that 78 percent reported that they had been celibate for varying periods of time: the majority for less than one year, but 35 percent for one to five years, and 8 percent for six years or more.

In 1994 *The Advocate,* a national gay and lesbian newsmagazine, reported the results of an in-magazine sex and health survey of male readers

(Lever, 1994). The men returned close to 13,000 questionnaires (out of 71,000 in circulation), but results are based on a systematic sample of 2,500. The majority (71 percent) described themselves as mostly or completely "out of the closet"; 97 percent described themselves as "gay" or homosexual and 3 percent as bisexual. Less than 1 percent reported they have never had sex with another man, whereas 4 percent reported that they had had no sex with another man in the past five years. (Among all men in the NHSLS study, 7.1 percent reported that they had been celibate for the past five years.) Another 12 percent reported that they had not had a sexual partner for two to five years, whereas 28 percent reported that they had been celibate for six months to two years.

In 1995 *The Advocate* conducted a similar survey of its female readers (Lever, 1995). The women returned more than 8,000 questionnaires, but results are based on a systematic sample of 2,513. The majority (62 percent) described themselves as mostly or completely "out of the closet": 89 percent described themselves as "gay," lesbian, or homosexual, and 9 percent as bisexual (2 percent are "not sure"). Slightly more than 2 percent reported that they had never had sex with another woman, whereas 4.6 percent reported that they had had no sex with another woman in the past five years. (Among all women in the NHSLS study, 8.7 percent reported that they had been celibate for that length of time.) Within the past five years, another 20 percent reported that they had not had a sexual partner for a period of two to five years, whereas 32 percent reported periods of celibacy lasting for six months to two years.

Summary

Although the studies cited above focus on behavior and not motivation or attitudes, we can tentatively suggest this summary: Some people who call themselves heterosexual, and who are predominantly heterosexual in behavior, also engage in homosexual acts. Some may experiment with homosexual behavior once or twice. Others may occasionally act on their attraction to people of the same sex, even if they call themselves heterosexual. Still others may recognize their attraction to others of the same gender, but they establish a heterosexual public persona and refrain from acting on these attractions or revealing their orientation to others. Finally, there are people who consider themselves to be homosexual or bisexual who, for whatever reasons (e.g., health concerns, religious convictions, or simply lack of opportunity), refrain from homosexual activities.

Prevalence of Proscribed Behaviors by Sexual Orientation

The sodomy provisions of the Uniform Code of Military Justice have been used as the basis for discharging homosexuals from the service. Some have argued that a policy allowing homosexuals to serve would be inconsistent with this provision of military law; however, unlike DoD Directive 1332.14, which prohibits same-gender partners regardless of sex act, Article 125 prohibits certain acts, *regardless* of gender of partner. Article 125 of the UCMJ states that a person engaging in "unnatural carnal copulation" with members of the same or opposite sex is guilty of sodomy. That is, under military law sodomy is forbidden whether performed by heterosexuals or homosexuals. The Manual for Courts-Martial (MCM) (DoD, 1994) defines sodomy as oral or anal sex (or sex with an animal). In this section, we review what is known about these forbidden behaviors in the general population. There are no published data on these behaviors among military personnel.

A review of available studies leads us to conclude that:

- oral sex, as defined and prohibited by the UCMJ/MCM, is widely practiced by both male and female homosexuals and by male and female heterosexuals;
- although a sizeable minority of heterosexuals has experienced anal sex at least once, most of them do not repeat this sexual act or else practice it infrequently—the majority of heterosexuals has not experienced this sexual act;
- although the prevalence of anal sex has decreased since the beginning of the AIDS epidemic, it is still a common sexual practice for many homosexual men.

Oral Sex among Heterosexuals and Homosexuals

In contrast to reports of same-sex behavior, reports of oral-genital sex should be less distorted by the problem of underreporting described above. Although this is a very private behavior, most Americans evidently consider it a "normal" sexual variation. For example, 88 percent of men and 87 percent of women in a large (albeit unrepresentative) national sample rated oral sex as "very normal" or "all right," versus "unusual" or "kinky." Even 77 percent of those who described themselves as "very religious" held this view (Janus and Janus, 1993).[13]

The NSM-1 (Billy et al., 1993), one of the probability samples described earlier, reports that among U.S. men between ages twenty and thirty-nine, 75 percent have performed and 79 percent have received oral sex. Among those currently married, 79 percent performed and 80 percent received it. Among the total sample, 32 percent of the men performed and 34 percent received oral sex within the last four weeks.

The NHSLS (Laumann et al., 1994) is the only study with a probability sample described in the table that provides comparable statistics on the prevalence of oral sex for both male and female respondents. Nearly three-quarters of men reported any lifetime incidence of fellatio, roughly the same proportion of women who reported any incidence of cunnilingus. Asked to describe their last sexual event, 27 percent of men said they had performed oral sex on their partner and 28 percent said they had received oral sex. The same proportion of women said they had received and performed oral sex during their last sexual event; however, roughly 8 percent fewer women than men reported oral sex during their last sexual event.

The NHSLS (Laumann et al., 1994) is also the only national probability sample study to offer comparable rates of lifetime experience with oral sex for homosexual men and women. Consider, however, that the data are based on very small numbers of respondents; only thirty-nine men and twenty-four women reported some level of homosexual or bisexual identity. Even though the number of women fell below the minimum of thirty that the authors set for computing group estimates, they found that 92 percent of women self-identified as homosexual or bisexual reported any lifetime incidence of receiving cunnilingus and 83 percent performed it (S. Michaels, personal communication, June 19, 1995). Of the men who self-identified as homosexual or bisexual, 89.5 percent had performed oral sex on a male partner and 89.5 percent had received oral sex from a male partner since puberty.

The RAND study described earlier is the only other study that we could find that included data on both heterosexual and homosexual respondents from a probability sample (Kanouse et al., 1991a, 1991b). Based solely on Los Angeles County residents, it is not generalizable to the U.S. population. RAND systematically sampled both homosexual and bisexual men and a random sample of the general adult male and female population in Los Angeles County.[14] Female homosexual respondents were not included. Among homosexual men who reported sex with another person in the past year, the proportion engaging in oral sex during the four-week period preceding the survey was 55 percent (Kanouse et al., 1991a). This proportion is about twice as large as the 26 percent of heterosexual men and women who report engaging in this behavior during the four-week period before the survey.

As magazine surveys, *The Advocate*'s studies of male and female sexuality are far from representative (Lever, 1994, 1995); both samples overrepresent those who are more highly educated and more affluent than the average American. We include the findings on oral sex here because the studies are recent, national in scope, and have the largest numbers of homosexual male and female respondents ever to participate in a sex survey. Among the male respondents to *The Advocate* survey, 85 percent received oral sex and 85 percent performed it during the past year. Among the female respondents, 73 percent received and 75 percent performed oral sex during the past year.

Although there are no published data on the prevalence of oral sex in a military population, it seems reasonable to assume, based on general population estimates from the studies reviewed above, that a majority of both married and unmarried military personnel engage in oral sexual activity, at least occasionally.

Anal Sex among Homosexuals and Heterosexuals

In contrast to reports of oral sex, reports of anal sex may share the same problem of underreporting described for same-gender sex. In Janus and Janus (1993), 71 percent of men and 76 percent of women rated anal sex as "unusual" or "kinky." These attitudes are in dramatic contrast to the same respondents' attitudes toward oral sex reported earlier, suggesting that anal sex is stigmatized behavior that is likely to be underreported.

The NSM-1 was the first study with a probability sample described in the table that included questions about the prevalence and incidence of anal sex (Billy et al., 1993). Reporting on men twenty to thirty-nine years of age—almost all of whom were heterosexual—20 percent reported that they had ever engaged in anal intercourse. But the percentage who have done so recently is much smaller; 90 percent of those who reported that they had ever had anal sex had not engaged in this sex practice in the four weeks prior to the interview. Younger men were less likely to have ever engaged in this sex practice: only 13 percent of those aged twenty to twenty-four compared to 27 percent of those aged thirty-five to thirty-nine reported that they had done so. Almost half of the small group of men who reported ever having had anal sex reported only one partner, whereas one out of five reported four or more partners.

In the NHSLS study (Laumann et al., 1994) 25.6 percent of the men in the total sample reported anal sex with a female partner, and 20.4 percent of the women reported anal sex during their lifetime. The heterosexual men most likely to have experienced anal sex in their lifetime were better-educated white men in their thirties and forties and Hispanic men. In contrast to lifetime experience, recent experience was significantly lower; only 9 percent of respondents reported heterosexual anal sex during the past year.

Lifetime experience with anal sex was dramatically higher among the small number of men in the NHSLS who identified themselves as gay or bisexual: 75.7 percent reported that they had been the insertive partner with a male and 81.6 percent reported that they had been the receptive partner.

The RAND study (Kanouse et al., 1991a, 1991b) provides the only comparative data on prevalence of anal sex among sizeable numbers of heterosexual men and women and homosexual men. In neighborhoods of Los Angeles County with large homosexual populations, a major epicenter of the AIDS epidemic, 34 percent of all homosexual/bisexual respondents who were sexually active in the year before the survey reported having engaged in anal sex with or without condoms during the four-week period immediately before the survey. This is more than six times the proportion (5 percent) of heterosexual men and women throughout Los Angeles County who reported engaging in this behavior during a comparable period. Homosexual respondents who described themselves as married to another male or in a monogamous primary relationship with another male were much more likely to report engaging in anal sex (58 percent versus 27 percent of all other sexually active homosexual respondents).

Other reports of the prevalence of anal intercourse among male homosexuals vary. We draw on two studies based on nonrepresentative samples that provide data on prevalence of anal intercourse, whether with or without condoms. Among the men who answered *The Advocate* survey, 58 percent reported that they had performed insertive anal intercourse and 56 percent reported that they had engaged in receptive anal intercourse in the past year (Lever, 1994). The second study suggests an age cohort difference in rates of anal intercourse. Among the 1,614 participants in the Pittsburgh Men's Study (Silvestre et al., 1993), 65 percent of homosexual men older than twenty-two reported anal receptive sex in the last two years, as did 81 percent of the men twenty-two years or younger. Anal insertive sex is reported by 78.5 percent of the older and 90 percent of the younger men in the 1992 study (A. J. Silvestre, personal communication, June 16, 1993).[15]

Data on behavioral changes in response to the increasing threat of AIDS from San Francisco, Chicago, New York City, and other large U.S. cities show that the prevalence of anal intercourse is affected by perceived risk of AIDS (Becker and Joseph, 1988; Stall, Coates, and Hoff, 1988). In the Pittsburgh study (Silvestre et al., 1993), the proportion who reported engaging in anal sex with at least half their partners declined from 45 percent in 1984 to 29 percent in 1988–1992. There is also some evidence suggesting that the incidence of this behavior, known as high-risk sexual activity for homosexual men, may be greater where there is low AIDS incidence (Turner, Miller, and Moses, 1989).

Great caution is needed in interpreting such disparate prevalence findings and in attempting to draw conclusions about average prevalence among all homosexual men. Data on homosexual men and women are necessarily based on samples of people who are willing to identify themselves as homosexual in orientation and/or behavior. Results from such samples cannot be taken as representative of the larger population that includes those unwilling to identify themselves. Moreover, as noted above, patterns of behavior—particularly engaging in anal sex—have undergone marked change in response to the AIDS epidemic. This indicates the prevalence data gathered a few years ago likely would not represent current behavior patterns. But change has not been uniform across geographic areas, so that the amount of change observed in one place cannot be incautiously applied to estimate change elsewhere.

Conclusion

In the past decade, our knowledge of the patterning and prevalence of sexual behaviors that are the subject of military regulations has increased dramatically. These data are relevant in any assessment of policy options open to the military. Because of the limitations of the data described at the outset of this chapter, we cannot offer precise answers to the questions framed in the introduction. Fortunately, precision is not needed to formulate the implications based on the data presented.

Data on the prevalence of homosexual behavior in the U.S. population suggest that the percentage of military personnel affected by the DoD Directive 1332.14 is small, even if occasional homosexual behavior, rather than regular homosexual behavior, is considered. The prevalence of predominantly or exclusively homosexual behavior in the U.S. population is undoubtedly higher than the 1 percent estimated from recent surveys using probability samples, but it is likely much lower than Kinsey et al.'s (1948) widely cited estimate of 10 percent. Data indicate that roughly 2 to 8 percent of adult American males acknowledge having had *any* sexual experience with another man during adulthood. Less is known about the prevalence of female homosexuality, but where data have been collected, estimates range from 1 to 6 percent who acknowledge having had sex with another woman during adulthood. Researchers cautiously report estimates as probable lower bounds of true prevalence inasmuch as stigmatized behaviors are underreported.

Although current military policy equates homosexual status (i.e., homosexual identity) with homosexual conduct (i.e., sexual behaviors), data indicate that they are not the same. A person who does *not* identify himself or herself as a homosexual may still engage in acts with someone of "the same sex for purposes of satisfying sexual desires" (in the language of DoD Directive 1332.14), whereas a person who *does* identify himself or herself as a homosexual may refrain from engaging in homosexual acts. Explicitly recognizing a distinction between status and conduct, the current policy gives commanding officers discretion to overlook a single instance of homosexual conduct. Data reviewed here indicate that exclusion from military service based on homosexual status alone would exclude some who do not engage in sexual acts with same-gender partners, while allowing others who do to serve.

Under Article 125 of the UCMJ, sodomy is forbidden—whether performed by heterosexuals or homosexuals. The Manual for Courts-Martial (DoD, 1994) defines sodomy as oral or anal sex, or sex with an animal. Because of the different assumptions made about the sexual practices of homosexuals and heterosexuals, Article 125 has been used to justify the exclusion of the former but not the latter. Insofar as penile-vaginal intercourse is, by definition, not an option for homosexual partners, it is assumed that their sexual conduct includes acts of sodomy. The conduct of heterosexuals, unless demonstrated otherwise, is not presumed to be in violation of military codes.

Our review of contemporary sexual behaviors shows that the assumption that heterosexuals are not engaging in sodomy is false. It is evident that the overwhelming majority of heterosexuals and homosexuals engage in some prohibited sex acts, primarily oral sex. A significant minority of heterosexuals and a majority of homosexual men engage in anal intercourse. If Article 125 were interpreted literally and enforced evenhandedly across the military population, such enforcement could affect at least three-quarters of all military personnel.

Notes

1. Convenience samples characterize most studies in both the sex research and epidemiologic literatures. Typically, samples are drawn from patients at sexually transmitted disease (STD) clinics, members of accessible organizations, persons who frequent public places for sexual contact, and volunteer respondents to magazine and other publicly announced surveys (Turner, Miller, and Moses, 1989). Contemporary researchers at the Kinsey Institute describe some of the other methodological shortcomings of sex research: small sample size, recruitment in one locale or just a few locales, and an overrepresentation of young, white, urban middle-class respondents (Reinisch, Sanders, and Ziemba-Davis, 1988).

2. One of the few studies related to this was conducted by Clark and Tifft (1966), who used a polygraph to motivate respondents (forty-five college males) to correct misreports they may have made in a previously completed questionnaire. They found that although 22.5 percent of these men ultimately reported some male-male sexual contact when confronted with a lie detector, only 7.5 percent of these had done so in the previously completed questionnaire.

3. The nature of Kinsey et al.'s sample may have affected the results: Some of the male subjects were prisoners, and there is reason to believe that the incidence of homosexual behavior is higher in prisons, as discussed below.

4. For females, Kinsey et al. (1953, pp. 473–474) reported that between 1 and 6 percent of unmarried and previously married females, but less than 1 percent of married females, were exclusively or predominantly homosexual by his definition—*in each of the years between twenty and thirty-five years of age*. They did not report an aggregate percentage for females regardless of marital status. But even if they had done so, the resulting percentage would not be comparable to the 10 percent for males because of differences in the age ranges and number of years required to qualify under the two definitions.

5. Kinsey, Pomeroy, and Martin (1948) estimate that two-and-a-half times as many men have been *predominantly or exclusively* homosexual as have been *exclusively* homosexual.

6. The National Survey of Men received a lot of attention in the popular press, in which it was more commonly referred to as the Battelle study and the Guttmacher study.

7. For example, Research Triangle Institute analysts estimate that there is a 95 percent probability that the "true" prevalence of Dallas men who engaged in

homosexual conduct in the previous twelve months is between 1.4 percent and 7.8 percent. This range is broad enough to include prevalence estimates in two of the three years for which GSS data have been reported.

8. Presented in the first two rows of the table.

9. Data from probability surveys are available for men only, but the same generalization can be made for women based on their higher separation rate for reasons of homosexuality in the U.S. military (GAO, 1992, p. 20).

10. The others were marital status (unmarried men were more likely to report same-gender contact); current religious affiliation (those with none were more likely to report same-gender contact); and size of city or town of current residence (those in places with populations of more than twenty-five thousand were more likely to report same-gender contact). The only social background variable associated with reports of same-gender contact was father's education: Respondents with college-educated fathers were more likely to report same-gender contact.

11. The researchers distributed questionnaires to a random sample of six hundred out of twenty-five hundred male prisoners in a medium-security prison; two hundred returned completed questionnaires, a 33 percent response rate. Because of the low response rate, we do not offer findings as estimates of prevalence; they are instructive, however, about the relationship between status and conduct.

12. In a counterpart study of the general population of Los Angeles County, Kanouse et al. (1991b) found that roughly 12 percent of the sample had been sexually inactive for five years or more.

13. The Janus Report, based on 2,765 volunteer respondents, is not representative of the U.S. population. We do not use it to draw conclusions about prevalence of behaviors, but we do draw on its data about sexual attitudes. Few general population surveys or epidemiologic studies measure attitudes toward particular sexual practices.

14. Data on some sexual practices, including both oral sex and anal sex, were obtained from subgroups that are not exactly comparable. Figures for heterosexuals represent everyone who had been sexually active in the previous five years, whereas those for homosexual men represent all those sexually active within the previous year.

15. This age difference in prevalence of anal sex is noted again in a report (Stall et al., 1992) on 401 randomly selected homosexual men who were interviewed by telephone in San Francisco in 1989: 23 percent reported having had *unprotected* anal sex in the past year. Forty-four percent of those aged eighteen to twenty-nine reported having had *unprotected* anal intercourse in the past year, compared with 18 percent of those thirty and older. This study is more typical of studies of risk behaviors in that it reports only on *unprotected* anal intercourse rather than rates with or without condoms.

References

Becker, M. H., & Joseph, J. (1988). AIDS and behavioral change to reduce risk: A review. *American Journal of Public Health, 78,* 394–410.

Billy, J. O. G., Tanfer, K., Grady, W. R., & Klepinger, D. H. (1993). The sexual behavior of men in the U.S. *Family Planning Perspectives, 25*, 52–60.

Blumstein, P. W., & Schwartz, P. (1976). Bisexuality in men. *Urban Life, 5*, 339–358.

Catania, J. (1995, November). *Effects of interviewer gender, interviewer choice, and item context on response to questions concerning sexual behavior.* Paper presented at the annual meeting of the Society for the Scientific Study of Sexuality, San Francisco.

Catania, J. A., Gibson, D. R., Chitwood, D. D., & Coates, T. J. (1990). Methodological problems in AIDS behavioral research: Influences on measurement error and participation bias in studies of sexual behavior. *Psychological Bulletin, 108*, 339–362.

Clark, J. P., & Tifft, L. L. (1966). Polygraph and interview validation of self-reported deviant behavior. *American Sociological Review, 31*, 516–523.

Committee on AIDS Research and the Behavioral, Social, and Statistical Sciences. (1990). Methodological issues in AIDS surveys. In H. G. Miller, C. F. Turner, & L. E. Moses (Eds.), *AIDS: The Second Decade,* (pp. 359–471), Washington DC: National Academy Press.

Davis, J. A., & Smith, T. W. (1991). *General Social Surveys, 1972–1991* [Machine-readable data file]. Chicago: National Opinion Research Center (Producer). Storrs, CT: The Roper Center for Public Opinion Research, University of Connecticut (Distributor).

Department of Defense (1982). *Enlisted administrative separations.* (Directive 1332.14). Washington, DC: U.S. Government Printing Office.

———. (1994). *Manual for courts-martial, United States.* Washington, DC: U.S. Government Printing Office.

Doll, L. S., Petersen, L. R., White, C. R., Johnson, E. S., Ward, J. W., & The Blood Donor Study Group. (1992). Homosexually and nonhomosexually identified men who have sex with men: A behavioral comparison. *The Journal of Sex Research, 29*, 1–14.

Fay, R. E., Turner, C. F., Klassen, A. D., & Gagnon, J. H. (1989). Prevalence and patterns of same-gender sexual contact among men. *Science, 243*, 338–348.

General Accounting Office. (1992, June). *Defense force management: DOD's policy on homosexuality.* (GAO/NSIAD-92-98), p. 20. Washington, DC: Author.

Janus, S. S., & Janus, C. L. (1993). *The Janus report on sexual behavior.* New York: Wiley.

Kanouse, D. E., Berry, S. H., Gorman, E. M., Yano, E. M., & Carson, S. (1991a). *Response to the AIDS epidemic: A survey of homosexual and bisexual men in Los Angeles County.* Santa Monica, CA: RAND Corporation.

Kanouse, D. E., Berry, S. H., Gorman, E. M., Yano, E. M., Carson, S., & Abrahamse, A. (1991b). *AIDS-related knowledge, attitudes, beliefs, and behaviors in Los Angeles County.* (RAND Report R-4054-LACH). Santa Monica, CA: RAND Corporation.

Kinsey, A. C., Pomeroy, W. B., & Martin, C. E. (1948). *Sexual behavior in the human male.* Philadelphia: W.B. Saunders.

Kinsey, A. C., Pomeroy, W. B., Martin, C. E., & Gebhard, P. H. (1953). *Sexual behavior in the human female.* Philadelphia: W.B. Saunders.

Laumann, E. O., Gagnon, J. H., Michael, R. T., & Michaels, S. (1994). *The social organization of sexuality: Sexual practices in the United States.* Chicago: University of Chicago Press.

Lever, J. (1994, August 23). The 1994 Advocate survey of sexuality and relationships: The men. *The Advocate,* pp. 16–24.

———. (1995, August 22). The 1995 Advocate survey of sexuality and relationships: The women. *The Advocate,* pp. 22–30.

Lever, J., Kanouse, D. E., Rogers, W. H., Carson, S., & Hertz, R. (1992). Behavior patterns and sexual identity of bisexual males. *The Journal of Sex Research, 29,* 141–167.

Loulan, J. (1988). Research on the sex practices of 1,566 lesbians and the clinical applications. *Women and Therapy, 7,* 221–234.

Miller, H. G., Turner, C. F., & Moses, L. E. (Eds.). (1990). *AIDS: The second decade.* Washington DC: National Academy Press.

Reinisch, J. M., Sanders, S. A., & Ziemba-Davis, M. (1988). The study of sexual behavior in relation to the transmission of human immunodeficiency virus: Caveats and recommendations. *American Psychologist, 93,* 921–927.

Rogers, S. M., & Turner, C. F. (1991). Male-male sexual contact in the U.S.A.: Findings from five sample surveys, 1970–1990. *The Journal of Sex Research, 28,* 491–519.

Silvestre, A. J., Kingsley, L. A., Wehman, P., Dappen, R., Ho, M., & Rinaldo, C. R. (1993). Changes in HIV rates and sexual behavior among homosexual men, 1984 to 1988/92. *American Journal of Public Health, 83,* 578–580.

Smith, T. W. (1992). Discrepancies between men and women in reporting number of sexual partners: A summary from four countries. *Social Biology, 39,* 203–211.

Sonenstein, F., Pleck, J., Ku, L., & Calhoun, C. *1988 National Survey of Adolescent Males* [Machine-readable data file and documentation, data set G6]. (1988). Washington, DC: The Urban Institute (Producer). Los Altos, CA: Sociometrics Corporation, Data Archive on Adolescent Pregnancy and Prevention (Producer and Distributor).

Stall, R., Barrett, D., Bye, L., Catania, J., Frutchey, C., Henne, J., Lemp, G., & Paul, J. (1992). A comparison of younger and older gay men's HIV risk-taking behaviors: The Communication Technologies 1989 cross-sectional survey. *Journal of Acquired Immune Deficiency Syndromes, 5,* 682–687.

Stall, R. D., Coates, T. J., & Hoff, C. (1988). Behavioral risk reduction for HIV infection among gay and bisexual men. *American Psychologist, 43,* 878–885.

Taylor, H. (1993, April 26). *The Harris Poll 1993 #20.* New York: Creators Syndicate, Inc.

Turner, C. F., Miller, H. G., & Moses, L. E. (Eds.). (1989). *AIDS, sexual behavior, and intravenous drug use.* Washington, DC: National Academy Press.

Valdiserri, R. O., Lyter, D. W., Leviton, L. C., Callahan, C. M., Kingsley, L. A., & Rinaldo, C. R. (1989). AIDS prevention in homosexual and bisexual men: Results of a randomized trial evaluating two risk reduction interventions. *AIDS, 3,* 21–26.

Wooden, W. S., & Parker, J. (1982). *Men behind bars: Sexual exploitation in prison.* New York: Plenum Press.

Sexual Orientation and the Military:
Some Legal Considerations

Peter D. Jacobson

On January 29, 1993, President Clinton directed the secretary of defense to draft an "Executive order ending discrimination on the basis of sexual orientation in determining who may serve in the Armed Forces of the United States" (Clinton, 1993). Following a series of discussions with congressional leaders, an interim policy, known as "Don't Ask, Don't Tell," was put in place, prohibiting the military from asking about a soldier's sexual orientation as long as the soldier did not refer publicly to his or her sexual orientation. The final policy, known commonly as "Don't Ask, Don't Tell, Don't Pursue," was announced on July 19, 1993, and subsequently enacted by Congress (10 U.S.C. §654).

This policy was intended to be more permissive than the ban on homosexuals in the military. Under the ban, homosexuality was deemed to be "incompatible with military service." The new policy softens that depiction somewhat by conceding that sexual orientation alone, absent homosexual conduct, is not grounds for discharge, though homosexual conduct remains a dischargeable offense. To many gay rights advocates, the change fell far short of the administration's commitment to end discrimination against homosexuals in the military. From a legal perspective, however, what is different is that the new policy accepts, in a limited

This chapter benefited from outstanding advice and counsel provided by Professor Stephen A. Saltzburg. I also appreciate the excellent advice provided by Bernie Rostker and Janet Lever throughout this project.

way, homosexual status. Under the ban, neither homosexual status nor conduct was acceptable.

As described by the administration, the intent of the new policy is to expand the "Don't Ask, Don't Tell" compromise to limit the circumstances under which a soldier can be investigated for homosexuality by creating a "zone of privacy" around the soldier's private sexual status. No questions will be asked about a soldier's sexual orientation, and homosexuals can serve in the military as long as they keep their sexual orientation to themselves. For example, the policy would allow participation in gay rights parades, dancing in gay bars, and reading homosexual materials, but would prohibit any outward display of homosexual conduct or statements of homosexuality. It is also designed to limit the grounds under which the military can investigate a soldier's sexual orientation or behavior, the "Don't Pursue" aspect of the new policy.

In essence, the policy codifies the distinction between status and conduct (as discussed below), except that it broadly defines conduct to include "a homosexual act, a statement by the service member that demonstrates a propensity or intent to engage in homosexual acts, or a homosexual marriage or attempted marriage." If credible information[1] about the soldier's sexual orientation were to be made public, or if a soldier states that he or she is homosexual, it would create a rebuttable presumption of homosexuality that could be rebutted with evidence of celibacy. If not rebutted, a discharge is likely. A statement by a soldier acknowledging homosexuality would create a rebuttable presumption of homosexual conduct or propensity to engage in such conduct.

How this policy will be implemented remains to be seen. An important legal question is whether courts will compel greater justification by the military for discharge based on sexual orientation alone (status) given the tacit admission that homosexuals are admitted and serve in the military. For the military, this means that its policies regarding accession and retention of homosexuals must be decided within the context of how the courts will respond to homosexual challenges to enter or remain in the military. The intense public scrutiny of the 1993 congressional hearings (e.g., Policy implications of lifting the ban, 1993) on homosexual service in the military ensures that the courts will be called upon to review the administration's policy. It is therefore crucial to assess how the courts might respond to the new policy.

Without much doubt, the legality, indeed the constitutionality, of the new policy will be resolved by the U.S. Supreme Court. Although it would be foolish to speculate on the outcome, I can identify the critical issues likely to be considered and provide an understanding of how the lower courts have resolved the litigation up to now.

This chapter examines the legal issues involved in adopting and implementing such a policy. I first consider the legal background, including legal and legislative trends regarding homosexuals and the current military policy toward homosexuals. I turn next to a discussion of general legal principles that are important for understanding how the courts have approached military cases, cases involving homosexual rights, and challenges to the ban on homosexuals in the military. Finally, I analyze the legal issues raised by the administration's policy.

Legal Background

For the past two decades, the courts, no less than society, have been engaged in determining the extent to which the Constitution of the United States protects homosexuals against discrimination. So far, homosexual rights advocates have had only limited success in the courts. Despite some notable court victories that we discuss below, particularly in adoption and family law, there is no discernible trend toward judicial recognition of homosexuality as a protected class. In particular, the Supreme Court's decision in *Bowers v. Hardwick* (1986), upholding the constitutionality of Georgia's sodomy statute, has been central to the political discussion of homosexual rights and has been a major legal barrier to the judicial expansion of homosexual rights.[2]

For example, challenges to laws or practices treating homosexuals differently from heterosexuals based on a denial of equal protection have foundered when courts have confronted the reality that *Bowers* permits homosexual conduct to be treated as a criminal offense. Thus, state laws and practices that treat homosexuals differently from heterosexuals have generally been upheld as long as states can show a rational basis for the differential treatment (the appendix contains a table that identifies which states currently have laws prohibiting sodomy). Because the majority culture tends to view homosexuality with anything from indifference to outright hostility, it is not surprising that courts have generally deferred to the state in challenges by homosexuals.

From the perspective of homosexual rights activists, however, the trend is probably viewed more propitiously. Starting from virtually no recognition twenty years ago, homosexuals have achieved victories on adoption and family matters that presage greater judicial success. This judicial success, coupled with generally limited legislative success, suggests that the courts will continue to be a primary battleground in society's struggle over homosexual rights and homosexual behavior. Perhaps more important, gay rights attorneys argue that the courts are now engaged in a broad dialogue about the extent of homosexuals' rights and roles within society. Although the result of this dialogue remains uncertain, that it is taking place at all is a major advance.

Aside from the political and policy questions regarding the "Don't Ask, Don't Tell, Don't Pursue" policy, several underlying legal issues have been raised by both proponents and opponents of the policy. First, and most important, will the courts overturn the policy (as a violation of the due process clause of the Fifth Amendment, insofar as it incorporates the equal protection clause of the Fourteenth Amendment), regardless of how it is implemented by the military? Second, what restrictions can legally be placed on homosexuals given the tacit acceptance of homosexual status? That is, will the courts sustain the distinction between status and conduct implied by the new policy? Third, if homosexuals were allowed to serve openly in the military, what privacy rights might be asserted by heterosexuals? And fourth, does the policy, unlike the ban that it replaces, unconstitutionally restrain freedom of speech in violation of the First Amendment? The answers depend on an analysis of recent trends in the law and whether homosexuals will be treated as a protected class for purposes of equal protection, a concept discussed in greater detail below.

Legal and Legislative Trends Regarding Homosexuals

As suggested above, recent trends regarding the protection of homosexuals from disparate treatment are mixed.[3] In areas such as family law and adoption, courts seem to be reducing barriers to homosexual participation. Recently, for instance, restrictions against homosexual adoptions have been overturned in several cases (e.g., *S.N.E. v. R.L.B.*, 1985; *In re Adoption of Charles B.*, 1990), homosexual couples have been recognized as a family in other cases (e.g., *Braschi v. Stahl Associates Co.*, 1989), and lesbians have been granted custody by several courts.

At a minimum, homosexual rights advocates argue that the courts are engaged in a dialogue about homosexual rights that is likely to result in expanded protections over time (W. Rubenstein and C. Feldblum, personal communication, April 1993). For example, recent court decisions in Nevada and Louisiana have struck down state sodomy laws as an unconstitutional invasion of privacy. And the Colorado Supreme Court recently struck down a voter initiative that would have overturned all local laws protecting homosexuals from discrimination as a violation of the equal protection clause of the Fourteenth Amendment (*Evans v. Romer*, 1994).

But other recent cases limit the scope of these rulings. In one case, the Virginia Supreme Court denied custody of a three-year-old to a mother living in a lesbian relationship, awarding custody to the maternal grandmother (*Bottoms v. Bottoms*, 1995). Perhaps more important, except in limited circumstances, advocates have generally not been successful in arguing that homosexuals should be treated as a protected class under the equal protection laws. In a challenge to a Cincinnati voter referendum that specifically denied antidiscrimination protections to homosexuals, the Sixth Circuit Court of Appeals held that the resulting amendment to the City Charter was constitutionally valid (*Equality Foundation v. City of Cincinnati*, 1995).

Whatever judicial success homosexuals have achieved has not been matched with corresponding legislative victories. For example, sexual orientation is not protected under the federal civil rights statutes, such as Title VII of the Civil Rights Act of 1964. Only nine states have enacted laws prohibiting discrimination based on sexual orientation, although several state legislatures are currently considering similar legislation. More than one hundred twenty municipalities have enacted similar ordinances. Although it is difficult to discern any trends at the state level regarding protections against discrimination, there appears to be a trend toward repealing or overturning sodomy statutes as applied to consenting adults. At the state level, homosexuals have had some success in repealing state sodomy statutes, although some twenty-three states still treat sodomy as a criminal offense (see appendix). During the 1994 elections, sponsors of antihomosexual ordinances (that is, ordinances similar to the Colorado voter initiative) were generally unsuccessful in obtaining voter support.

Military Policy Regarding Homosexuals
For many years, military policy regarding accession and retention of homosexuals was based on Department of Defense Directive 1332.14 (DoD, 1982),[4] which stated that, "Homosexuality is incompatible with military service. The presence in the military environment of persons who engage

in homosexual conduct or who, by their statements demonstrate a propensity to engage in homosexual conduct, seriously impairs the accomplishment of the military mission. . . . As used in this section: (1) Homosexual means a person, regardless of sex, who engages in, desires to engage in, or intends to engage in homosexual acts; . . . (3) A homosexual act means bodily contact, actively undertaken or passively permitted, between members of the same sex for the purpose of satisfying sexual desires."

Under this directive, any soldier who acknowledged his or her homosexuality or whose sexual orientation was discovered (through, for instance, an investigation or statement by someone else) was subject to being discharged from the military. The policy made no distinction between status and conduct; a soldier could be discharged either for being a homosexual or for engaging in a homosexual act. If the military determined that a homosexual encounter was a onetime experience (such as a heterosexual engaging in a homosexual act) or a departure from the soldier's usual and customary behavior (such as resulting from intoxication), adverse action need not be taken automatically.

An important aspect of both the ban and the current policy regulating homosexuals is Article 125 of the Uniform Code of Military Justice (10 U.S.C.A. §925). Under Article 125, any person who engages in unnatural carnal copulation (defined in the Manual for Courts-Martial [DoD, 1984] as oral or anal sex) with another person of the same or opposite sex or with an animal is guilty of sodomy and subject to a court martial. As written, Article 125 applies equally to homosexuals and to heterosexuals.[5] Allegations of unequal treatment notwithstanding, available data on prosecutions under Article 125 show that both heterosexuals and homosexuals have been prosecuted for sodomy. But the reach of Article 125 goes beyond that captured in the prosecution statistics (Burrelli, 1993). As a practical matter, most homosexuals facing an Article 125 charge are given the option of an administrative discharge (based on honorable conditions) instead of standing trial. There is currently no exclusion in the Manual for Courts-Martial pertaining to Article 125 for private, consensual sex between adults.[6]

General Legal Principles

As indicated by the lengthy testimony presented by several legal scholars at Nunn's 1993 hearings on the administration's proposed policy changes, there are numerous legal issues presented by reconsidering the ban that could be discussed in this chapter.[7] Because this chapter is limited to the legal consequences of policy adopted, the range of legal issues is narrowed somewhat. Nevertheless, it is important to discuss some general legal principles pertaining to homosexuals in the military before considering the legal implications of the current policy.

Deference to the Military

Perhaps one of the strongest doctrines in the law is that the courts generally defer to the military on matters relating to military service, organization, and personnel. As the U.S. Supreme Court stated in *Rostker v. Goldberg* (1981), "Judicial deference . . . is at its apogee when legislative action under the congressional authority to raise and support armies and make rules and regulations for their governance is challenged." This broad deference has a long history that is premised on the understanding that military service is fundamentally different from civilian life. It is thus generally accepted that persons entering the military give up certain constitutional rights and have fewer privacy expectations than in civilian life.

Given that premise, the courts are reluctant to second-guess the needs of the military based largely on principles derived from and applicable to civilian society. Policies determined by the military and for the military are generally treated with great deference, even when the restrictions would otherwise be unconstitutional within a civilian context. In *Goldman v. Weinberger* (1981), for example, the Court refused to uphold a challenge by an Orthodox Jew to a restriction that prohibited him from wearing a yarmulke when in uniform. Although such a restriction in a civilian setting would violate the First Amendment, the Court held that the needs of the military for good order and discipline, as well as sameness of appearance, superseded Goldman's right to wear what was admittedly an unobtrusive skull cap. It is also interesting to note that Congress subsequently enacted specific legislation to overturn the *Goldman* decision.

As a general principle, therefore, any policy option considered by the secretary of defense starts with what amounts to a presumption of constitutional validity. In effect, this allows the military great discretion in accession and retention policies (the issues of most interest to us right now), including setting the conditions under which individuals may enter and serve in the military. Consequently, courts have upheld restrictions as to age, height, weight, single parentage, previous drug use or criminal conviction, and so forth that might not survive scrutiny under civilian circumstances. That is, the military may set conditions that discriminate against various groups. Those challenging military rules and policies have the burden of proving that the rule or regulation does not serve a rational military interest. As numerous court cases have shown, that is a difficult burden to overcome.

Equal Protection and the Military

One way to overcome the burden of deference to the military is to challenge the regulation as a violation of equal protection of the laws based on membership in a protected class, such as a racial minority. Being recognized as a protected class is important because of the level of scrutiny that the courts will therefore apply to a governmental rule or regulation. To shorten what would otherwise be a lengthy discussion of a somewhat convoluted area of jurisprudence, equal protection applies if the regulation contravenes a fundamental right, such as the right to privacy, or if the group subject to disparate treatment constitutes a protected class. If homosexual sodomy were to be considered a fundamental right of privacy, laws making such behavior a criminal offense would be unconstitutional. But in *Bowers,* the Court held that homosexual sodomy does not constitute a fundamental right, and so upheld laws making sodomy a criminal offense.

Technically, because *Bowers* was a due process challenge, some scholars have argued that the result does not preclude a finding that homosexuals should be treated as a suspect class for an equal protection challenge (e.g., Sunstein, 1988). Most courts, however, have held that homosexuals cannot be a protected class when such an important activity as sexual conduct can be treated as a criminal offense. It is important to add that even if *Bowers* were overturned, this would not definitively answer the question of whether open homosexuals could serve in the military, although it might undermine the policy reasons for upholding the current policy. The issue of what kinds of homosexual conduct are disruptive and can be subject to sanctions would remain. In *Goldman,* for instance, the plaintiff could not be punished for being Jewish, but could be punished for wearing a yarmulke in violation of the regulations.

As an alternative to reliance on fundamental rights, homosexuals can use the equal protection laws to challenge the validity of the ban. Over time, courts have developed three levels for judging a governmental regulation's validity under the equal protection laws. First, strict scrutiny will be applied to classifications, such as race, that are inherently suspect. Any regulation of a suspect class must serve a compelling state interest and be narrowly tailored to meet that interest. Very few regulations can survive this test.[8] Second, intermediate or heightened scrutiny will be applied to classifications, such as gender, that are usually invalid but for which some justification can be presented. Under heightened scrutiny, any regulation must be substantially related to an important governmental interest. Third, when no suspect class is determined, the regulation will be reviewed on a rational basis test. This test presumes the validity of governmental regulation as long as the classification is rationally related to a legitimate state interest. Under passive rational basis, courts generally sustain the regulation, most typically economic and social legislation, so long as it serves any reasonable state interest. Under active rational basis, an emerging doctrine, courts will require additional justification for any restrictions. Just what level of proof is required to satisfy active rational basis is not clear at this point.

So far, federal courts have not treated homosexuals as a suspect class for equal protection. As a consequence, challenges to the validity of military policies by homosexuals will be decided on the rational basis test. In the past, the passive rational basis test has been applied when considering deference to the military. Recently, some lower federal courts have begun to apply the active rational basis test to military cases. If that trend continues, the sustainability of the current policy will depend on what level of justification is needed to satisfy the active rational basis test. To sustain the current policy, or to impose certain restrictions against homosexuals in the military, would require the military to go considerably beyond the rationales offered to date in support of these policies. On the other hand, if courts adopt a passive rational basis standard for equal protection analysis, the military will have little problem in defeating legal challenges.

The congressional hearings (e.g., Policy implications of lifting the ban, 1993) and the RAND findings (National Defense Research Institute, 1993) generated during the national debate on the Clinton administration's policy might play an important role in determining how courts respond to the military's justification for its policies toward homosexuals. The extensive empirical work provided for the secretary of defense could form the basis both for any restrictions imposed against homosexuals and for defining a coherent rationale that can be defended in court. Of course, the courts could just as easily determine that the agreement between Congress and the administration indicates a genuine policy concern that open homosexuality undermines morale and unit cohesion.

Responding to the Prejudices of Others

Two Supreme Court cases, *Palmore v. Sidoti* (1984), and *City of Cleburne v. Cleburne Living Center, Inc.* (1985), have held that private biases and potential injuries resulting from those prejudices are insufficient grounds for policy determinations.[9] As the Court stated in *Palmore* (p. 433), "The Constitution cannot control such prejudices but neither can it tolerate them. Private biases may be outside the reach of the law, but the law cannot, directly or indirectly, give them effect." At this point, it is uncertain how this principle will be applied in the context of a homosexual challenge to certain restrictive military policies.

The reality of military cases such as *Goldman* is that the military can regulate what members do or say precisely because certain actions are likely to disrupt morale and undermine unit cohesiveness. No one could ban wearing a yarmulke in civilian life; yet it could be banned in the military as emphasizing individual differences over group identity. And the *Palmore* and *City of Cleburne* cases may have little bearing on military regulations that rest upon a judgment that certain behaviors are immoral.

Even so, the Ninth Circuit, in *Pruitt v. Cheney* (1991), required the government to prove on the record that Pruitt's discharge did not rest on the prejudice and bias of other soldiers against homosexuals. The court specifically stated that the military's justification would be examined in light of *Palmore* and *City of Cleburne*. In the context of an active rational basis analysis, a court might use the *Palmore* principle to negate previously accepted reasons or justifications for adopting a particular restriction. As a result, military policy made on the basis that some people are uncomfortable with homosexuals might not survive a *Palmore/Cleburne* challenge, absent an independent rationale.

What Privacy Rights Can Heterosexuals Assert?

An important policy consideration is to balance the privacy rights of members of the military who object to homosexuality with the principle that homosexual status is acceptable to the military. Through flexible command policy, privacy concerns could be alleviated by ensuring freedom from personal and sexual harassment and maximizing flexibility in sleeping and bathroom facilities, where feasible. As a legal matter, however, there appear to be few ways in which a heterosexual could assert a privacy right sufficient to bar adoption of the current policy.

For one thing, it is generally understood that a soldier yields full privacy rights upon entering the military. For another, courts would be likely to balance individual privacy rights with the opportunities of others to serve in the military. Courts may well rule in an individual case that the assertion of a privacy right is sufficiently compelling to justify rescinding the contract between the soldier and the military (that is, to allow an early discharge). And an individual commander might attempt to accommodate an individual soldier who had deep moral objections against rooming with a homosexual. But courts would be unlikely to override the military's policy choice to allow homosexuals to serve based on heterosexual soldiers' privacy rights. This would be especially true if courts were to treat homosexuals as a protected class.

Just as important, granting a privacy right to heterosexuals who object to serving with homosexuals must be justified on grounds other than private biases or prejudices against homosexuals. As discussed above, the *Palmore* and *City of Cleburne* cases send a strong message that the courts will not sanction policies based on private biases. Thus, it might be difficult to construct a general heterosexual privacy right that satisfies the *Palmore/Cleburne* test.

Freedom of Speech Issues

Homosexuals have not had much success in using the First Amendment to challenge the ban. For instance, First Amendment challenges to the ban on status have failed. Although courts have held that soldiers may discuss homosexuality, read homosexual materials, and even advocate a change in policy, the courts have held that there is no right of expression in the military to state "I'm gay," no right of free association to join homosexual organizations, and no right just to be homosexual (see, e.g., *Ben-Shalom v. Marsh,* 1989; *Pruitt v. Cheney,* 1991). These cases were decided in the context of a ban on open homosexuality. Under the current policy, the definition of what constitutes conduct, including the circumstances and consequences of the action, would determine the outcome.

It is possible, as opponents of the current policy argue, that the policy makes a First Amendment challenge more plausible because the policy punishes the content of speech—for instance, just saying "I'm gay." As Williams (1994) points out, the effect of the policy is to deter speech, regardless of whether conduct is impermissible. More problematic is that the policy is aimed at the content of the speech. Traditionally, restrictions on speech that are content-neutral, such as restrictions on numbers of pickets, are more likely to be upheld than restrictions that impede the expressive content of the speech. In particular, when the speech is directed at matters of public concern, courts are less likely to uphold restrictions. Thus, a factual issue under the current policy will be whether a soldier's admission of homosexual status constitutes a statement of personal concern or a matter of public importance. If it is the former, the policy is more likely to be sustained.

In other words, the legal issue will be whether the administration can define status so narrowly (and conduct so broadly) that statements of personal sexual orientation are not protected speech under the First Amendment. As an example of this difficulty, suppose that acknowledged homosexuality was acceptable, but any homosexual conduct was unacceptable. In some cases, the distinction between telling (probably protected conduct) and doing (prohibited conduct) becomes very difficult to determine. Suppose a soldier states that he has engaged in anal sex while a member of the armed forces. Is this telling or doing? Is this grounds for an investigation or discharge? In the context of the current policy, these questions may receive more attention than they did under the ban. Even the court in *Cammermeyer v. Aspin* (1994), a case otherwise ruling in favor of a challenge to the ban, rejected a First Amendment challenge, noting that the plaintiff was discharged not because of speech but because of sexual orientation (see, e.g., *Ben-Shalom v. Marsh,* 1989; *Goldman v. Weinberger,* 1981; *Pruitt v. Cheney,* 1991). But, as noted below, the only reported case to have considered the current policy upheld a First Amendment challenge to restrictions on admitting homosexual status.

Homosexuals in the Military: Current State of the Law
Given the above legal principles, it is not surprising that most challenges by homosexuals to the ban have been unsuccessful. It should be noted, however, that most of the reported cases were litigated under the previous policy banning homosexuals from serving in the military. The implicit acceptance of homosexual status under the new policy means that these cases cannot be directly applied to the "Don't Ask, Don't Tell, Don't Pursue" regime, although the ways courts resolve the status-conduct distinction will be an important determinant of how cases under the new policy are decided. At present, there is only one reported case challenging the current policy. As noted below, it is likely that this case will be decided by the U.S. Supreme Court.

Except for cases brought in the Ninth Circuit Court of Appeals, few challenges have succeeded. And no successful district court case has survived an appeal outside of the Ninth Circuit. No appellate court—even in the Ninth Circuit, where the ban has been under sustained attack—has ruled that restricting homosexual conduct is unconstitutional or has held that homosexuals are a suspect or quasi-suspect class for equal protection challenges. Two cases recently decided by the Ninth Circuit and the D.C. Circuit presented clear equal protection challenges to the ban, but neither court, as discussed below, decided the case on equal protection grounds (*Meinhold v. United States Department of Defense,* 1994; *Steffan v. Perry,* 1994).

A typical case ruling against challenges to the previous ban on homosexuality is *Dronenburg v. Zech* (1994), in which the D.C. Circuit held that the Navy's policy of mandatory discharge for homosexual conduct violated neither the equal protection laws nor the sailor's right to privacy. Most significantly, the court stated that any change in the ban should be made by elected officials, not by the courts. Taking basically a passive rational basis approach, the court added that, "The effects of homosexual conduct within a naval or military unit are almost certain to be harmful to morale and discipline. The Navy is not required to produce social science data or the results of controlled experiments to prove what common sense and common experience demonstrate" (p. 1398).

What may be changing, however, is the standard of review for justifying the military's ban on homosexuals. At least in the Ninth Circuit, the standard has already shifted to an active rational basis analysis. Relying on *Pruitt,* the district court in *Meinhold* explicitly rejected deference to military judgment as a rationale for discharging homosexuals. If followed in other cases, this would subject the ban or other restrictions against homosexuals to greater judicial scrutiny by forcing the military to justify any restrictions.

Recent district court cases. Recent cases suggest that some district courts are more aggressively confronting the military's rationale for opposing homosexuals than the circuit courts. In *Cammermeyer v. Aspin* (1994) and *Dahl v. Secretary of the United States Navy* (1994), the district courts confronted the equal protection questions directly, both holding that the military had not met its burden of justifying the ban on gays and lesbians, but neither court provided protected status for homosexuals. (Both cases were decided under the military's prior policy of banning homosexuals.) Following the lower court's decision in *Meinhold,* these courts relied on the *Palmore* and *City of Cleburne* rationale to determine that the ban was based solely on prejudice toward homosexuals. Holding that the ban amounted to prejudice and could not withstand active rational basis analysis, the courts rejected the military's justification that the ban ensured unit cohesion and morale.

Just as important, these cases rebuffed the administration's arguments that two previous cases, *United States v. Harding* (1992) and *Heller v. Doe* (1993), had rejected the active rational basis standard in favor of passive rational basis. Both the *Cammermeyer* and *Dahl* courts reasoned that the *Harding* and *Heller* cases meant only that the government need not present any evidence to justify its actions. The court retains the obligation to inquire into the prejudicial effects of a policy when the plaintiff presents evidence challenging a policy's rational basis. In short, courts are not obliged to accept the military's stated rationale for excluding homosexuals without examining the underlying motivations of the policy.

Recent circuit court cases. The circuit courts, however, have been more circumspect in recent decisions. Two recent decisions reached opposite conclusions, but neither court attempted to consider independently whether the policy was an inherent violation of equal protection.

For example, in *Steffan,* the D.C. Circuit, sitting *en banc,* overruled a three-judge panel that had declared the ban to be unconstitutional under an active rational basis standard. In an opinion that does not confront the constitutional questions directly, the court *en banc* upheld Steffan's discharge from the Naval Academy based on the connection between Steffan's admitted homosexual status and presumed propensity to commit homosexual acts. In essence, the court adopted the reasoning of *Ben-Shalom,* discussed below, and held that "a statement that one is a homosexual can rationally be used by the Navy as a proxy for homosexual conduct—past, present, and future." Because Steffan conceded that the regulations regarding homosexual conduct were valid, the court's ruling that status equals conduct foreclosed his appeal. Much of the opinion, however, is a discussion of whether Steffan's refusal to answer conduct-related questions amounted to an admission of prohibited conduct.

In *Meinhold,* the Ninth Circuit also focused on the status-conduct distinction, but held that status alone could not be equated with conduct, and hence could not be penalized. Like the *Steffan* decision, the court assumed that homosexual conduct could be prohibited (*Steffan v. Perry,* 1994). And, like the *Steffan* case, the court refused to rule directly on

Meinhold's claim that gays should be a protected class. Instead, the court ruled that it could not be rationally presumed that homosexuals would violate regulations prohibiting homosexual conduct. Although the *Meinhold* court did not address whether the military's policy rationale amounted to prejudice, the court noted that "Nothing in the policy states that the presence of persons who say they are gay impairs the military mission. Rather, the focus is on prohibited conduct and persons likely to engage in prohibited conduct" (p. 1479). As a result, a statement that "I'm gay" manifests no concrete propensity to commit prohibited acts.

Challenging the current policy. As noted above, only one case challenging the "Don't Ask, Don't Tell, Don't Pursue" policy has been reported. In *Able v. Perry* (1995), plaintiffs challenged the current policy as a violation of their First and Fifth Amendment rights. Each of the six named plaintiffs admitted their status as homosexuals, but denied any actual homosexual conduct or propensity to engage in such behavior.

Based primarily on an analysis of the First Amendment's right to freedom of speech, the court enjoined the military from discharging the plaintiffs. In some of the strongest language seen in any of the reported cases on homosexuality, the court held clearly that an acknowledgment of a member's sexual status is protected speech: "To presume from a person's status that he or she will commit undesirable acts is an extreme measure. Hitler taught the world what could happen when the government began to target people not for what they had done but because of their status" (p. 974). The court went on to hold that the current policy cannot be defended by the military's desire to accommodate the privacy interests of heterosexual service members. Such a policy rationale, according to the court, amounts to a policy based on private prejudice, which is unconstitutionally discriminatory. The Clinton administration is appealing the case. In all likelihood, this case will eventually be decided by the U.S. Supreme Court.

Status versus conduct. What has emerged in most of the recent cases as a major point of contention is the distinction between homosexual status (sexual orientation alone) and homosexual conduct (engaging in homosexual acts). How the status/conduct issue is framed and decided is crucial to understanding the courts' responses to challenges to the military's policy on homosexuals. In general, courts have consistently held in nonmilitary cases that individuals may be sanctioned for their conduct, but not as a result of their status (Williams, 1994). For example, in *Robinson v. California* (1962), the U.S. Supreme Court held that an individual could not be convicted of a criminal offense for being addicted to a narcotic substance.

A reading of the recent cases noted above suggests that attorneys for homosexual soldiers have agreed (perhaps for tactical reasons) that a soldier can be discharged for homosexual conduct while on duty.[10] In *Meinhold,* the court stated that, "On the merits, we defer to the Navy's judgment that the presence of persons who engage in homosexual conduct, or who demonstrate a propensity to engage in homosexual conduct by their statements, impairs the accomplishment of the military mission" (p. 1472). Focusing on status, the plaintiff's attorneys argue that the status of being a homosexual, absent demonstrated conduct, is insufficient to justify discharge. Supporters of either the ban or the current policy argue instead that it is ludicrous to separate status from conduct in determining a soldier's propensity to commit a homosexual act. But this position conveniently ignores surveys showing that at least 75 percent of heterosexuals engage in oral sex, conduct also prohibited under Article 125 (see chapter 2).

Except for *Able,* each of the recent cases considering the military's policy on homosexuals has stated that the military can justify banning homosexual conduct. The issue, therefore, is whether the military is justified in assuming that homosexual orientation is tantamount to homosexual behavior (that is, being a homosexual predisposes the person toward homosexual conduct). Courts following *Cammermeyer* and *Meinhold,* in ruling against the military, have determined that status and conduct are distinct and that the military cannot discharge someone merely for the status of homosexual orientation. Courts following *Steffan* in rejecting the status-conduct distinction have upheld the military even if the military fails to show any actual homosexual conduct.

Although perhaps difficult to defend, courts could uphold the status-conduct distinction as a rational policy choice. Yet, the courts might rule that once acknowledged homosexual status was permitted in the military, an absolute ban on sexual conduct could not be maintained. At issue is whether the policy choice to distinguish between status and a particular form of sexual conduct would be a rational one based on military considerations, and hence acceptable under the deference to the military principle, or whether the distinction could not be defended as a rational means of achieving a policy goal. The legal argument against its constitutionality would be that the premise of the distinction, that status is separable from conduct, is internally inconsistent and unsustainable. When confronted directly by that contradiction, courts would rule that once status is acknowledged, a ban on conduct violates equal protection.[11]

To the best of our knowledge, there are no cases holding that policies based on the status-conduct distinction are unconstitutional, and there are cases explicitly rejecting this position as applied to homosexuals.[12] Although numerous courts discuss the salience of the distinction, no court has ruled that recognizing homosexual status requires equating status with conduct, or that recognizing status requires a change in policy regarding conduct. And some courts have ordered the reinstatement of acknowledged homosexuals without questioning the ban on homosexual conduct (*Watkins v. United States Army,* 1989). In most areas of the law, what is prohibited is certain conduct, not the status of the actor. For the argument equating status and conduct to be tenable, a court must equate status with conduct as a matter of law, something that few courts have done in the past.

Nevertheless, there is language in *Ben-Shalom* suggesting that the distinction between status and conduct defies common sense. The court stated that, "Plaintiff's lesbian acknowledgment, if not an admission of its practice, at least can rationally and reasonably be viewed as reliable evidence of a desire and propensity to engage in homosexual conduct. . . . [I]t is compelling evidence that plaintiff has in the past and is likely to again engage in such conduct. . . . The Army need not shut its eyes to the practical realities of this situation" (p. 464). And *Steffan* can certainly be read expansively as suggesting that the distinction is untenable, at least in a military context.

In *Ben-Shalom,* the Seventh Circuit ruled that the ban on status could remain because the admission of status is tantamount to an admission of conduct. Because the court determined that status amounted to conduct, the ban on conduct could be enforced. The court ruled that the regulation was constitutional because the Army did not need to ignore the connections between status and conduct. Thus, the court based its ruling on the prohibited conduct, regardless of whether status is banned.

Nothing in the *Ben-Shalom* and *Steffan* cases, however, requires the military to adopt a policy that equates status with conduct. In other words, neither case holds that the military cannot base its policy on distinguishing status from conduct and penalizing only the latter. Just because the current policy acknowledges status does not mean that a court will therefore rule that the military cannot continue to ban sodomy where the military chooses as a policy not to equate status with conduct. The acknowledgment of sexual orientation need not have an effect on how the military enforces its ban on sodomy. Although this means that the tension between the current policy and Article 125 is not unconstitutional, it does not mean that the tension disappears. To see why, imagine the issue in a heterosexual context. Given that it is likely that most married heterosexual couples engage in oral sex, an act prohibited by Article 125, should they be presumed to be in violation of Article 125 simply because of their marital status (see chapter 2)? If the answer is "no," the argument that status alone constitutes a violation of Article 125, and hence mandates the unconstitutionality of the current policy, must fail.

There are three additional reasons for this conclusion. First, the *Ben-Shalom* and *Steffan* courts ruled only on whether the military could equate status with conduct, not whether it was mandated, as a matter of policy, to do so. As noted in the *Meinhold* decision, the military can legitimately determine that disruption to good order and discipline emanates from sodomy, and that restricting sexual conduct rather than status is a legitimate policy objective.

Second, as noted above, the weight of the cases is that policy choices made by the military will be given great deference by the courts. As long as the policies are not irrational, courts are likely to defer to military judgment. At one point, the *Ben-Shalom* court stated flatly, "If a change of Army policy is to be made, we should leave it to those more familiar with military matters than are judges not selected on the basis of military knowledge" (*Ben-Shalom v. Marsh*, 1989, p. 461). Even if the distinction between status and conduct is artificial, the administration's policy would start with the presumption of validity based on deference to the military. This remains a difficult standard to overcome.

Third, the lower court applied a heightened scrutiny analysis after holding that homosexuals constituted a suspect class. As discussed above, relatively few courts have so held, and the appellate court in *Ben-Shalom* explicitly rejected this finding. In all likelihood, a challenge to the current policy would still, at most, be judged under an active rational basis test. The use of the rational basis test, when combined with traditional deference to military policy, suggests that the military should be able to defend its policy choice of acknowledging status while prohibiting sodomy, especially if it treated heterosexual sodomy in a similar manner.

Toward the Supreme Court

From this review, it appears that the Supreme Court is likely to face four issues in determining whether to uphold the administration's policy. First, are gays entitled to protected status under equal protection? Second, is the distincton between status and conduct sustainable, or must the military presume conduct from status? Third, is the military's rationale for excluding homosexuals based on prejudice? Fourth, does the "Don't Ask, Don't Tell, Don't Pursue" policy amount to an unconstitutional infringement of the right to freedom of speech?

Despite the current state of the law, there are now some lower court decisions and some powerful dissents, including Justice Blackmun's dissent in *Bowers* and Judge William A. Norris's dissent in *Watkins*, that could provide a road map for overturning the "Don't Ask, Don't Tell, Don't Pursue" policy. As shown in *Evans,* the Colorado Supreme Court's recent decision restricting the ability of voters to overturn ordinances protecting homosexuals from discrimination, the litigation context is dynamic.[13] More to the point, Judge Eugene H. Nickerson's opinion in *Able* directly challenges the policy's inherent free speech infringement.

But even if the judiciary becomes more rigorous in applying equal protection for homosexuals in civilian cases, the question still remains how far judges will go in scrutinizing military regulations. As the circuit court opinions in *Meinhold* and *Steffan* suggest, absent homosexuals being classified as a suspect class, deference to the military remains a powerful judicial doctrine. This is an issue likely to be decided by the U.S. Supreme Court, however. Even if the Court is unprepared to revisit *Bowers,* it can still determine that the current policy toward homosexuals is motivated by prejudice, or is an unconstitutional restraint on freedom of speech. If the Court wants to overturn the policy, there are many options available short of including gays as a protected class. If the Court wants to sustain the policy, it need only rule that the standard of review is passive rational basis, virtually accepting the military's stated rationale.

Notes

1. Credible information is defined as information that "supports a reasonable belief that a service member has engaged in such conduct. It requires a determination based on articulable facts, not just a belief or speculation."

2. Technically, *Bowers* was a due process challenge. Some scholars have ar-

gued that the result does not preclude a finding that homosexuals should be treated as a suspect class for an equal protection challenge. But most courts have held that homosexuals cannot be a protected class when such an important activity as sexual conduct can be treated as criminal.

3. For an exhaustive review of trends in legislation and case law, see Developments in the Law (1989). For an excellent, and more recent compendium, see Rubenstein (1993, especially pp. xv–xxi). In the private sector, homosexuals have had some success in obtaining domestic partnership benefits from large corporations, such as AT&T and Microsoft.

4. This directive was promulgated in 1981. Although the ban on homosexuals predates DoD Directive 1332.14, previous policy permitted the retention of open homosexuals at the military's discretion. The directive was issued in response to numerous court challenges, such as *Matlovich v. Secretary of the Air Force* (1978), questioning why some open homosexuals were discharged although others were retained. The 1981 directive removed the military's discretion in deciding whether to retain an open homosexual, making such discharge mandatory.

5. Historically, state sodomy statutes have been widely perceived as being the legal basis to exclude or punish homosexuality (see, e.g., *Bowers v. Hardwick*, 1986, dissenting opinion by Justice Blackmun). Even where the statutes are sex-orientation neutral, they have not been enforced equally against homosexual and heterosexual behavior. In the military, there are indications that Article 125 has been used differentially for homosexuals and heterosexuals. Threats to homosexuals of prosecutions under Article 125 have been used to elicit confessions of homosexuality and then acceptance of administrative discharges. The current policy makes the consequences of unequal enforcement more serious: Homosexuals who practice oral or anal sex would be exposed to the risk of court-martial proceedings without the availability of an administrative discharge as an option.

6. Jacobson (1993, pp. 358–362) specified an administrative mechanism whereby the administration could limit the scope of Article 125 by exempting consensual sex between adults from its coverage. Because this option was not adopted by the administration, it is not discussed in this chapter.

7. For excellent discussions of the broad range of potential legal issues, see the testimony by Stephen A. Saltzburg, David A. Schlueter, and David F. Burrelli, presented to the Senate Armed Services Committee Hearings, March 1993.

8. Becoming a protected class, however, is easier said than done. Courts have applied three principal criteria to determine whether a particular class should be protected under the equal protection laws: (1) history of discrimination against a discrete group; (2) classification based on immutable or distinguishing characteristics, and (3) lack of political power. No federal appellate court has held that homosexuals meet these criteria, although courts differ on which aspects are not satisfied.

9. *Palmore* was a racial discrimination case, so it might not be broadly applied; *Cleburne* was a zoning regulation and was applied to a particular set of facts.

10. This argument has the short-term tactical advantage of focusing a court's attention on the soldier's ability to serve actively in the military, particularly given the courts' traditional reluctance to sanction status. But it has the potential long-

term disadvantage of justifying disparate treatment of homosexual soldiers. It may also lead to endless arguments over what constitutes conduct absent a more explicit code of conduct suggested by Jacobson (1993).

11. Arguably, the Ninth Circuit is the appellate court most likely to seize on these arguments to overturn the ban altogether. As noted above, some judges on the Ninth Circuit would like to overturn the ban even if the military makes no policy changes. But even Judge Norris's dissent in *Watkins v. United States Army* (1989), one of the strongest statements opposing the military's ban on homosexuals, focuses on sexual orientation without making the connection to conduct presumed by this argument.

12. See, for example, *Pruitt v. Cheney* (1991) and *Meinhold v. United States Department of Defense* (1993). In *Steffan v. Cheney* (1990), the court rejected the government's argument that Steffan's sexual orientation created a rebuttable presumption that he had committed homosexual acts. See also *Jacobson v. United States* (1992), in which the court stated that "evidence that merely indicates a generic inclination to act within a broad range, not all of which is criminal, is of little probative value in establishing predisposition. . . . Furthermore, a person's inclinations and 'fantasies . . . are his own and beyond the reach of government' " (pp. 1541–1542).

13. See also *Baehr v. Lewin* (1993), holding that Hawaii's ban on same-sex marriages constitutes gender-based discrimination and is suspect under Hawaii's constitutional equal protection provision.

References

Able v. Perry, 800 F. Supp. 968 (E.D.N.Y. 1995).

Baehr v. Lewin, 852 P.2d 44 (Haw. 1993).

Ben-Shalom v. Marsh, 881 F.2d 454 (7th Cir. 1989).

Bottoms v. Bottoms, No. 941166, 1995 Va. LEXIS 43 (Va. Apr. 25, 1995).

Bowers v. Hardwick, 478 U.S. 186 (1986).

Braschi v. Stahl Associates Co., 543 N.E.2d 49 (N.Y. Ct. App. 1989).

Burrelli, D. F. (1993). *Homosexuals and U.S. military policy.* CRS report for Congress. (Report No. 93-52F). Washington, DC: Congressional Research Service.

———. Statement of Dr. David F. Burrelli Before the Senate Armed Services Committee Hearings, March 29, 1993.

Cammermeyer v. Aspin, 850 F. Supp. 910 (W.D. Wa. 1994).

City of Cleburne v. Cleburne Living Center, Inc., 473 U.S. 432 (1985).

Clinton, W. J. (1993, January 29). Memorandum to the Secretary of Defense: "Ending Discrimination on the Basis of Sexual Orientation in the Armed Forces."

Dahl v. Secretary of the United States Navy, 830 F. Supp. 1319 (E.D. Cal. 1994).

Department of Defense. (1982). *Enlisted administrative separations.* (Directive 1332.14). Washington, DC: U.S. Government Printing Office.

———. (1984). *Manual for courts-martial, United States.* Washington, DC: U.S.

Government Printing Office.

Developments in the law: Sexual orientation and the law. (1989). *Harvard Law Review, 102,* 1508–1671.

Dronenburg v. Zech, 741 F.2d 1388 (D.C. Cir. 1984).

Equality Foundation v. City of Cincinnati, 1995 U.S. App. LEXIS 10462 (6th Cir. 1995).

Evans v. Romer, 854 P.2d 1270 (Colo. 1994).

Goldman v. Weinberger, 475 U.S. 503 (1981).

Heller v. Doe, 113 S. Ct. 2637 (1993).

Hunter, N. D., Michaelson, S. E., & Stoddard, T. B. (1992). *The rights of lesbians and gay men: The basic ACLU guide to a gay person's rights* (3rd ed.). Carbondale, IL: Southern Illinois University Press.

In re Adoption of Charles B., 522 N.E.2d 884 (Ohio 1990).

Jacobson v. United States, 503 U.S. 540 (1992).

Jacobson, P. D. (1993). Sexual orientation and the military: Some legal considerations. In National Defense Research Institute, *Sexual orientation and U.S. military personnel policy: Options and assessment* (pp. 332–367). Santa Monica, CA: RAND.

Matlovich v. Secretary of the Air Force, 591 F.2d 852 (D.C. Cir. 1978).

Meinhold v. United States Department of Defense, 808 F. Supp. 1453 (C.D. Cal. 1992), aff'd, 34 F.3d 1469 (9th Cir. 1994).

National Defense Research Institute (1993). *Sexual orientation and U.S. military personnel policy: Options and assessment.* Santa Monica, CA: RAND.

Palmore v. Sidoti, 466 U.S. 429 (1984).

Policy implications of lifting the ban on homosexuals in the military: Hearings before the Committee on Armed Services, House of Representatives, 103d Cong., 1st Sess. 1 (1993). Washington, DC: U.S. Government Printing Office.

Pruitt v. Cheney, 963 F.2d 1160 (9th Cir. 1991).

Robinson v. California, 370 U.S. 660 (1962).

Rostker v. Goldberg, 453 U.S. 57 (1981).

Rubenstein, W. B. (Ed.). (1993). *Lesbians, gay men, and the law.* New York: New Press.

Saltzburg, S. A. Statement Before the Senate Armed Services Committee Hearings, March 29, 1993.

Schlueter, D. A. Statement Before the Senate Armed Services Committee Hearings, July 22, 1993.

S.N.E. v. R.L.B., 699 P.2d 875 (Alaska 1985).

Steffan v. Cheney, 920 F.2d 74 (D.C. Cir. 1990).

Steffan v. Perry, 41 F.3d 677 (D.C. Cir. 1994).

Sunstein, C. R. (1988). Sexual orientation and the constitution: A note on the relationship between due process and equal protection. *University of Chicago Law Review, 55,* 1161–1179.

Uniform Code of Military Justice, 10 U.S.C.A. §§801 et seq.

United States v. Harding, 971 F.2d 410 (9th Cir. 1992).

Watkins v. United States Army, 875 F.2d 699 (9th Cir. 1989).

Williams, K. (1994). Gays in the military: The legal issues. *University of San Francisco Law Review, 28,* 919–955.

Appendix

Current Status of Sodomy Restrictions, by State

Sodomy Restrictions	No Sodomy Restrictions
Alabama	Alaska
Arizona	California
Arkansas*	Colorado
Florida	Connecticut
Georgia	Delaware
Idaho	Hawaii
Kansas*	Illinois
Louisiana	Indiana
Maryland	Iowa
Massachusetts**	Kentucky
Michigan	Nebraska
Minnesota**	Nevada
Missouri*	New Hampshire
Montana*	New Jersey
North Carolina	New Mexico
Oklahoma*	North Dakota
Rhode Island	Ohio
South Carolina	Oregon
Tennessee*	Pennsylvania
Texas	South Dakota
Utah	Vermont
Virginia	Washington
	Washington, D.C.
	West Virginia
	Wisconsin
	Wyoming

Sources include Hunter et al. (1992); personal communications: Thomas F. Coleman, Executive Director, Spectrum Institute, Los Angeles, CA; Jon Davidson, American Civil Liberties Union (ACLU), Los Angeles Office; Professor Arthur Leonard, New York Law School, New York; William B. Rubenstein, ACLU New York Office; and applicable case law.

*Restriction applies to same-gender sex only.

**Sodomy laws remain in force, but states ban discrimination on the basis of sexual orientation.

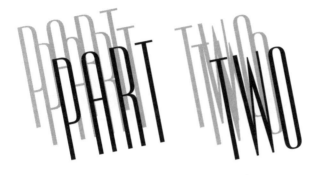

PART TWO

Relevant Experience
from Other Domains

Integration of Women in the Military: Parallels to the Progress of Homosexuals?

Patricia J. Thomas and Marie D. Thomas

Women's struggle to achieve equity in the military may foreshadow the integration of homosexuals. Military women historically have been viewed as a renewable resource to be utilized in time of need and returned to civilian life when the crisis passed. In a similar manner, the sexual orientation of homosexuals has been overlooked during wartime and deemed intolerable in peacetime so that they, too, could be discharged (Shilts, 1993).

The civil rights movement provided a model for discriminated-against groups to follow, yet women and homosexuals fit uneasily into the role. Numerically, they are minorities in the armed forces and have experienced discrimination. In suits against the Department of Defense (DoD), however, women and homosexuals have encountered courts reluctant to intrude upon the authority of the military in matters of national defense (Binkin and Bach, 1977; Sarbin, 1991).

The status of women in the military changed profoundly during the 1970s. Tracing the source of unequal treatment of military women and the national climate that allowed change to occur may result in a greater understanding of the controversy surrounding homosexuals.

The views expressed in this chapter are those of the authors, are not official, and do not necessarily reflect the positions of the U.S. Navy or the Department of Defense.

Women's Integration in the Military

Legislation enacted during the presidency of Harry Truman regulated the integration of women into the military following World War II. It was not the intent of Congress to treat women the same as men, but to provide for their continuous presence in the military. As a consequence, the Women's Armed Services Integration Act of 1948 contained many restrictive clauses (Thomas, 1981). Disparate treatment was not only legal, but mandated.

Other legislation banning Navy and Air Force women from assignment to aircraft or vessels with combat missions was inserted into 10 U.S.C. §§6015 and 8549. The assignments of Army and Marine Corps women were controlled by service policies, which were equally restrictive.

The participation of homosexuals in the armed forces was, until recently, not based in federal law, but constrained through military regulation. In 1919, when the Articles of War were revised, sodomy became a felony, giving a behavioral definition to what constituted homosexuality. During World War II, the first regulations designed to exclude from military service those with "homosexual tendencies" were written. Because homosexuals technically were not members of the armed forces, no regulations managing their participation were required.

The 1970s: A Decade of Change

Pressures impinging upon the military during the 1970s were responsible for the movement toward greater integration of women. An internal impetus was military need. When the draft expired, military recruiting fell on hard times. The all-volunteer force was initiated, but young men were reluctant to affiliate with the armed services. Manpower requirements were high because of the Vietnam War, and the number of recruitable men was falling along with the birthrate. Women became an untapped personnel resource, but some constraints had to be removed so that they could be more fully utilized.

A second impetus for change was women's willingness to use the courts to achieve greater equity in the military. This newfound activism resulted from two social movements—civil rights and feminism. African American civilians had demonstrated that disparate treatment could be changed through litigation and the feminist movement encouraged women to challenge gender-specific laws and regulations.

Civilian leaders in the DoD and members of Congress also brought pressure on uniformed military leaders to modify policies restricting women or be faced with more extreme change. For example, Senator William Proxmire (D-WI) introduced legislation in 1977 to abolish the sections of the federal code that prevented women from flying combat aircraft and sailing aboard all Navy ships. The election of Jimmy Carter to the presidency in 1976 resulted in appointees to DoD who were more sympathetic to treating women servicemembers equally than their Republican predecessors had been.

Military need during the 1970s did not result in modifications to regulations against homosexuals. Shilts (1993) has argued, however, that during the Vietnam War (and in other wars) known homosexuals were tolerated and in peacetime, homosexuals in critical jobs were overlooked. Social movements, the courts, and politics did affect the military's treatment of homosexuals, but not until much later.

The 1980s: Backlash

Although there had always been opponents to the increased role of women in the military, the first hint of a serious halt in the progress occurred in 1981. Interestingly, 1981 was also the year in which the DoD issued its first policy on homosexuality. In February, the Army announced a "pause" in its five-year plan to increase the number of women from 61,000 to 87,500 by 1985. An eighteen-month internal study began that resulted in recoding twenty-three military occupational specialties as combat-related and classifying others as having very heavy physical demands. Thus, the policies developed during "womanpause" implied that half of the women in the Army were either unable, or would not be allowed, to perform in the jobs they held.

The integration of women was derided as a "social experiment" (Cropsey, 1980). Retrenchment began as it became apparent that the Equal Rights Amendment would not be ratified by three-fourths of the states. One of the most effective moves by opponents of women in the military, discussed by Stiehm (1989), was to redirect the debate away from equal opportunity and focus on readiness and effectiveness. Women's inferior upper-body strength, pregnancy, single parenthood, and "physical problems" were cited as sufficient justification to restrict their roles in the military (Holm, 1982). These arguments reached their nadir with the publication of Brian Mitchell's book, *The Weak Link: Women in the Military,* in 1989. Mitchell argued that women had profoundly weakened the armed forces by adversely affecting morale and readiness.

Also symptomatic of the backlash were the witch-hunts that targeted groups of women for discharge on the basis of homosexuality. The investigations were often initiated by disgruntled men whose advances had been rebuffed (Benecke and Dodge, 1990). The most notorious case was the investigation of sixty-five drill instructors from the Marine Corps Recruit Training Depot at Parris Island, South Carolina. The Navy's defining incident was the attempt to discharge nineteen of the sixty-one women aboard the USS *Norton Sound* for homosexuality (only two were discharged). All eight of the black women in the crew were among the accused. The Army also hunted lesbians during this period, discharging eight military police officers from West Point and investigating the women's softball team at the academy. Women athletes were particularly suspect, and the volleyball team at the Air Force Academy was also investigated (Shilts, 1993).

The backlash against women was manifested as harassment, both sexual- and gender-based. Aboard ship, women officers' quarters were trashed; women at the academies were repeatedly told that they did not belong and were referred to by unflattering epithets; overseas, they were subjected to social isolation. A survey conducted by the DoD found that 64 percent of military women reported being sexually harassed in 1989 (Martindale, 1991), though it was not known whether this rate reflected an increase over previous years.

The 1990s: Resurgence

The 1990s began auspiciously for proponents of equal rights and responsibilities for military women. In the fall of that year, Operation Desert Shield was launched and women deployed with their units. When the Persian Gulf War erupted, reservists of both sexes were ordered into action. Women were taken prisoner, killed in action, and were casualties of a Scud missile attack. The media brought into America's living rooms the reality that military women were performing many of the same jobs and exposed to the same danger as men.

President Bush created the fifteen-member Commission on the Assignment of Women in the Armed Forces (referred to as the Presidential Commission throughout this chapter) in December 1991 to review the laws and service regulations that barred women from combat. The commission's final report recommended removing all restrictions on shipboard assignments for Navy women (except for submarines), restoring the ban on Air Force and Navy women in combat aircraft, and continuing the

Army and Marine Corps policies of restricting women from ground combat.

In 1992 Bill Clinton campaigned for president on a platform that favored equal rights for women and opposed the military's ban on homosexuals. Democratic appointees in the DoD ignored recommendations of the Presidential Commission that would restrict the assignments of military women and accepted those that expanded assignments. In November 1993, Congress repealed the section of the United States Code that barred Navy women from combat ships. In January 1994, the secretary of defense rescinded the rule that barred women from noncombat units based on an assessment of risk of hostile fire.

But incidents of egregious sexual harassment continued to haunt the military. Perhaps the most celebrated case was the physical attacks against 80 women at the 1991 Tailhook Convention of Navy and Marine Corps pilots. Other women came forward in mid-1994 to testify before Congress about physical assault, pressure for sexual favors, and their frustration over the barriers to obtaining relief from sexual harassment.

The General Accounting Office (GAO), at the request of the Senate, reviewed sexual harassment at the military academies (GAO, 1994). Despite almost twenty years of integration, 50 percent of women midshipmen at the Naval Academy, 76 percent of women cadets at the Military Academy (West Point), and 59 percent of the women at the Air Force Academy stated that they experienced harassment at least twice a month. Most of this harassment was based on gender, consisting of derogatory comments, statements about women having lowered standards, and offensive graffiti or gestures.

Women and Gays in the Military

Masculine Ideology

Women are on the fast track in becoming naval carrier aviators. As time passes and they become more numerous . . . the macho males who now make up the groups will be leaving in droves.

Also, when a thousand or more women make a six- to nine-month cruise with married men, the wives back home will be asking their husbands to get out of the Navy. They will not go for a Navy Carnival cruise line.

Then what? Women may well be fighting the next war while

the men remain at home. (Letter to the editor, *San Diego Union-Tribune,* November 29, 1994, p. B7)

The writer of the letter expresses a fear that increased numbers of women will lead to a feminized military—a military in which "macho men" will not want to participate. His attitude supports Shilts's (1993) claim that "[t]he issue of women in the military was never about women; it was about men and their need to define their masculinity" (p. 492). For when women can do "a man's job," that job can no longer be used as a sign of manhood.

The armed forces have represented a passage from boyhood to manhood for generations of American men. As such, the services define what it means to be a man. Karst (1991), in his article on racial desegregation of the military, describes as "masculine ideology" an ideal of manhood that is a rejection of the feminine. His assertion that femininity is viewed as "a fundamental flaw—a failure, at the deepest level, to qualify" (Karst, 1991, p. 504), is well supported by certain military traditions. Common insults hurled at recruits in basic training refer to them as girls, pussies, or chickens (Rustad, 1982). Marching cadences subordinate women or emphasize male sexual power over them (DePauw, 1988). The purpose is to inculcate the belief that only *real* men are fit to defend the country. Perhaps Marine Corps General Robert Barrow said it best in explaining his objection to women in combat: "When you get right down to it, you have to protect the manliness of war" (quoted in Wright, 1982, p. 74).

The ideology of masculinity is also about power, and the military is the ultimate symbol of American might. "The heart of the ideology is the belief that power rightfully belongs to the masculine—that is, to those who display the traits traditionally called masculine" (Karst, 1991, p. 505). For decades, women were denied positions of power within the military. Women were permitted to serve in support functions, but not to carry guns, shoot cannons, or launch torpedoes. Because of these limitations, women could never aspire to the highest ranks within the military because they had no experience in the primary purpose of the military—war.

A major function of the military, therefore, is to create men—men defined by a traditional masculine ideology. This ideology requires that "real" men repress the feminine inside them. Gay men, however, "[call] into question everything that manhood is supposed to mean" (Shilts, 1993, p. 5). According to Shilts, limiting the roles of both women and gay men served the same purpose—to defend traditional masculine ideology in a world that is rapidly changing.

Separating the Experiences of Lesbians and Gay Men

There is a tendency, when discussing issues related to gays in the military, to lump together the experiences of lesbians and gay men. This inclination reflects a traditional approach in our society to understanding women's experience through a male paradigm (Meyers, 1994). When considering arguments against gays in the military, however, it becomes clear that such arguments often refer to gay men and not to lesbians. For example, Meyers (1994), in her discussion of *Washington Post* coverage of the "repeal the gay ban" controversy, refers to the argument that openly gay personnel would undermine the "unit cohesion" needed for fighting. She points out that this argument illustrates the invisibility of women in general and lesbians in particular, because the discussion is about disrupting *male* bonding—the absence of women is presumed. (This same argument has often been used as a rationale for not allowing women into combat.) She reports that women were all but invisible in *Post* articles covering the controversy: "[B]y excluding women, the military appeared overwhelmingly male, bound by 'threads of male camaraderie' " (Meyers, 1994, p. 341).

In fact, hostility within the military toward lesbians on the one hand, and gay men on the other hand represents different (but related) aspects of the fight to preserve the military as a bastion of masculinity. Women (obviously) and gay men are not *real men* in the way that manhood has traditionally been defined. Lesbians, in the military context, are thought of as women first. In many cases, sexual orientation has not been the reason for their harassment but was used as a *tool* to harass them. The persecution of gay men in the military, on the other hand, has everything to do with their sexual orientation.

Recognizing that the experiences of lesbians and gay men in the military are not the same, we divide the rest of the chapter into two sections. First, we consider how attitudes toward lesbians in the military have been entwined with attitudes toward women in general. The second section focuses on women, gay men, and the strikingly similar arguments against their participation in combat.

Lesbians and the Integration of Women

According to a recent GAO report (1992), during fiscal years 1980 through 1990, women were discharged for homosexuality at a rate consistently higher than their representation in each branch of the military. Although representing only 10 percent of military personnel, women constituted 23 percent of those discharged for homosexuality (GAO, 1992).

There is no way of knowing what percentage of women in the military are lesbians. According to Shilts (1993), the percentage might have been as high as 80 percent during World War II, dropping to 35 percent in the early 1980s, and currently standing at about 25 percent. Early criteria for women in the military were especially stringent and would have eliminated a large pool of heterosexual women. For example, married women were not allowed to enlist and pregnancy resulted in an automatic and immediate discharge. During the 1960s, when the women's movement began to make nontraditional jobs attractive to a broader spectrum of women, the percentage of lesbians in the military probably dropped dramatically (Shilts, 1993).

Persecution of lesbians involves more than sexual orientation; it has represented the military's opposition toward all women. As Shilts (1993) said, "The profound victimization of lesbians in the military has less to do with homophobia than with sexual discrimination and harassment, the kind faced by women breaking into occupations once reserved for men" (p. 5). We are not saying that lesbian and heterosexual women have had exactly the same experiences in the military. Lesbians always have been in greater danger of persecution than have heterosexual women. But heterosexuality has not protected women from being threatened, investigated, and forced out of the service.

The witch-hunts so graphically detailed by Benecke and Dodge (1990) and Shilts (1993) seem to be one end of a continuum of sexual harassment of women, especially women entering nontraditional occupations or job sites. This extreme harassment of women is not new:

> From the first entry of women into specialties other than nursing through the Women's Auxiliaries of World War II, purges have occurred in a pattern which seems to be linked to women's entry into nontraditional job fields, personnel gluts which provide a rationale for decreasing the numbers of women in the military, and the social and political climate of the time. (Benecke and Dodge, 1990, p. 218)

It is important to emphasize that hostility toward women in the military increased when women entered nontraditional jobs. Women who were in "feminine" occupations, such as nurses and clerk-typists, were accepted. Traditional female occupations, however, are support functions; they are not viewed as directly contributing to the military's mission—national defense. A nontraditional job is, by definition, one that had been the exclusive territory of men and is performed primarily by men. Such a job is often spoken of as "masculine." When women enter these occupational fields, the job's function as a marker of masculinity is reduced or destroyed. One way to avoid "demasculinization" is to make the women who take on these jobs deviant, *not* women; that is, to label them "lesbian" (Benecke and Dodge, 1990). (Segregating women in this way also protects military men's economic and status interests; see Benecke and Dodge, 1990.) This label potentially carries a heavy cost: "any time a woman is called a 'dyke,' her reputation, her career, and even her liberty are on the line" (Benecke and Dodge, 1990).

Military women in nontraditional jobs, therefore, are living a paradox. They must learn to be feminine enough to avoid charges of being a lesbian. If they are too feminine, however, they will be viewed as incompetent and inferior (Benecke and Dodge, 1990).

Another issue of importance revolves around women's sexual accessibility to men. According to Karst (1991), the characteristics of a "real" man are to be "sexually aggressive and yet protective toward women" (p. 504). Military women who choose to limit their accessibility to men challenge masculine ideology. Lesbian-baiting has been triggered by a woman's refusal of a man's sexual advances (Benecke and Dodge, 1990); according to DePauw (1988), anecdotal evidence suggests that threatening to denounce a woman as homosexual is *the* most prevalent form of sexual harassment in the military.

Obviously, such harassment takes a toll on military women. The psychological and physical effects of sexual harassment are well documented, including such symptoms as anxiety, depression, headaches, sleeping disorders, and weight loss (Gutek and Koss, 1993; Terpstra and Baker, 1991; Thomas, 1994). Other effects reach beyond individual women. DePauw (1988) claims that fear of investigation for lesbianism "is perhaps the fundamental reason why the Navy has had trouble filling its seagoing billets and why all the services have trouble recruiting women for non-traditional careers" (p. 41). Women may give in to unwelcome sexual advances to avoid being labeled as lesbians (Benecke and Dodge, 1990). Shilts (1993) reports that military women have become pregnant in advance of investigations for homosexuality. Finally, fearing that too much contact will generate charges of homosexuality, military women in nontraditional jobs remain isolated from each other. Although they are few in number, they cannot seek assistance or comfort from each other or teach each other how to get through the military successfully (Benecke and Dodge, 1990; DePauw, 1988). It seems plausible, then, that lifting the homosexual ban would benefit *all* military women by removing the stigma (and danger) associated with the label "lesbian."

Women, Gay Men, and Combat

Combat, either defensive or offensive, is the raison d'être of the military. Excluding women from combat marginalizes their role and ensures that they cannot aspire to the highest positions within the services. Assignment policies for women in the Navy and Air Force were, and still are in the Army and Marine Corps, shaped by combat exclusion legislation or service regulation. In a democracy dedicated to equality under the law, the only legitimate rationale for excluding women from combat is that they would jeopardize the military mission.

The arguments against women in combat are remarkably similar to those against gay men in the armed forces. The policy statement supporting the incompatibility of homosexuality and military service appears in DoD Directive 1332.14, 1982 (see chap. 1, pp. 6–7). It maintains that the presence of homosexuals would adversely affect good order and discipline, trust among servicemembers, integrity of rank, right to privacy, recruiting, public opinion, and security. We discuss the points raised in the directive as they have been argued to exclude women and how they similarly have been applied to gay men.

Maintenance of discipline, good order, and morale. This argument against women in combat focuses on sexuality. It posits that men will compete for (i.e., fight over) the favors of women, losers will experience some degree of unhappiness, and the morale of the entire unit will suffer. It is axiomatic that disciplinary problems would increase in such a scenario. Thus, women must be excluded or men will misbehave. James Webb (1979) raised this argument to its ultimate extreme when he discussed intraservice rape as a rationale for not permitting women in combat.

When applied to gay men, the disciplinary problem is believed to be threefold: (1) homophobic men would aggress against gays; (2) aggressive gay men would prey upon young men who are insecure in their sexuality; and (3) gay men without partners would act out their jealousy. Violence against men who are known or suspected to be gay has certainly occurred in the military (Shilts, 1993) and is an issue of concern, but viewing aggression against gays as a "disciplinary problem" places blame on the victim rather than on the perpetrator of the violence. The other two disciplinary concerns assume that removal of the ban would cause gay men (and lesbians) to lead openly homosexual lives. In fact, removal of the ban would not necessarily make the military a "safe" environment for homosexuals, and it is possible that many would remain closeted.

We believe that part of the maintenance-of-good-order argument against gays in the service is relevant, at least in the short term. If the ban were lifted, antihomosexual violence would increase because the targets of violence would be more easily identified than when they were closeted. But these hate crimes would decrease if the military took prompt, unequivocal action against the perpetrators. The other part of the argument, which posits that aggression by gays would increase disciplinary problems, is not credible. There is no evidence to suggest that gay men would be any more predatory or likely to act out their jealously than the heterosexual men in the military.

Foster mutual trust and confidence. This argument often includes the concepts of bonding and unit cohesion. The military deeply believes that cohesion is an essential psychological component of an effective fighting unit. The cohesion argument has been used in the past to exclude African Americans (see chapter 5) and is being used today to exclude gay men and women. With women, the problem may arise from sexual competition.

> When you get a bunch of guys out in the field, a kind of closeness develops. I guess you'd call it camaraderie. They come to be a group. They work together. It really is a bonding process. Put women with them, and it's instantly every man for himself. They're all plotting how to [form a more perfect union with] the ladies. It doesn't matter whether the women are interested or not. Cohesion is gone. (Reed, 1994)

Women's effect on these constructs is difficult to assess because the sustained stress, deprivation, and fear experienced in actual combat is almost impossible to simulate. Yet Colonel Paul Roush (1990), who served with the Marine Corps in Vietnam, noted that there sometimes was a bonding problem between blacks and whites in his unit. When enemy firing began, however, bonding became very "doable."

Trust and confidence in a military setting also refer to being totally dependent on each other for completion of the mission and, perhaps, survival. Because women (and gay men) are characterized as lacking sufficient aggression, physical strength, and stamina for combat, they cannot be trusted to come to the aid of another servicemember. Because straight men would not accept gay men, men would not bond and unit cohesion would suffer.

Homosexuals who subscribe to military values (as do most who join the military), work toward group goals, and do not look or act "different" probably would not disrupt bonding. But the presence of homosexuals who, through their behavior or appearance, are a constant reminder of alternative values could negatively affect unit cohesion. Personnel with antipathy toward gays and lesbians would minimize their interactions with them at work and during leisure hours. If other personnel supported the homosexual servicemember, the atmosphere would become divisive.

Ensure the integrity of rank and command. Regulations against fraternization existed before growth in the number of military women occurred. The scenario envisioned was that of a superior unwilling to order a junior into a life-threatening situation because of friendship between the two. Now, fraternization has come to signify a sexual bond between a junior woman and a senior man. The breakdown of command is no longer the reluctance to endanger the junior, but favoritism in more mundane situations. Abuse of authority is another aspect of deterioration of the integrity of rank. Male superiors using their status to gain favors from female subordinates is at the core of quid pro quo sexual harassment. Male subordinates cannot compete in this game.

Integrity of rank is also believed to be threatened when women become leaders. Men may refuse to assume the role of follower because taking orders from a woman infringes upon traditional masculine ideology. According to Gilder (1986), effective leaders embody the values of their followers. How can a straight woman or a gay man embody male values?

This argument similarly invokes the specter of gay men (and lesbians) using their rank to gain sexual favors from subordinates or giving favors to a superior in return for some benefit. We fail to see why the incidence of such occurrences would be any greater among homosexual personnel than among heterosexual personnel. Moreover, the belief that gay leaders would be unable to command the respect and obedience of straight men contradicts history. Alexander the Great and Frederick II of Prussia were acknowledged homosexuals who led victorious armies. The effectiveness of a leader is far more important to his followers than his sexual orientation.

Privacy. When this argument is applied to women, it usually focuses on cost. Because our society does not condone the mixing of the sexes in bedrooms or bathrooms except within the privacy of the home, military facilities that were formerly all-male have to be reconfigured or facilities must be built to accommodate women. Ships have to undergo costly overhaul to install toilet bowls (instead of urinals) in heads and bulkheads (solid walls) within berthing areas. Overshadowing this concern is the scenario of ground combat. What will women do when nature calls? How will they change their tampons when menstruating?

Straight men's repulsion of gay men's ogling their bodies in communal showers or of unwelcome advances is at the heart of this argument when applied to homosexuals (see chapter 11 for a more complete discussion of this issue). Men are women's predators but are discomforted by the thought of becoming the object of the chase.

The privacy argument is very weak. As pointed out by RAND researchers (National Defense Research Institute, 1993), servicemembers relinquish full privacy when they join the armed forces. Moreover, because public accommodations make no distinction based on sexual preference, the courts would not be likely to rule in favor of a heterosexual who felt his privacy rights were abridged by having to live in close quarters with homosexuals.

Recruitment and retention. Since the mid-1970s, the DoD has expressed concern that if women were assigned to combat positions, men's propensity to affiliate with the military would decline (Borack, 1978). One of the issues addressed in a recent survey of servicemembers was the potential impact on recruiting and retention of assigning women to direct combat (Roper, 1992). Only 18 percent of all active-duty men in the sample thought the services' ability to attract high school graduates with high aptitude test scores would be negatively affected by such a change, and only 19 percent said it would decrease their likelihood of remaining in the service.

Apparently, as illustrated in the letter to the editor that began this section, the fear is that if women infringe too deeply into the traditional male role, men will no longer see the military as a masculine institution. Or the problem may be that if women can do it, it is not worth doing.

Homosexual warriors may also threaten the military's image as a forger of men. As mentioned earlier in the chapter, the military is a mechanism by which traditional masculine ideology is imparted to men in our culture. If, as Karst (1991) says, the characteristics of a "real" man include dual responses to women—sexual aggression and protection—gay men fail the test because they are not interested in women sexually. In addition, part of traditional masculinity is a repression of the "feminine" within oneself (Karst, 1991). The stereotype of gay men is that they do *not* repress their femininity and may actively display it. A military made up of women and gay men would represent the antithesis of masculine ideology and, therefore, would be an unattractive career alternative to those who believe the military should embody traditional masculinity and the power associated with it.

It is difficult to gauge the extent to which admitting homosexuals would negatively affect recruitment and retention. Canada and Australia, which have lifted their bans on homosexuals, have experienced no change in enlistments or reenlistments. Because the United States shares a common culture with these nations, a similar outcome would logically follow.

Maintenance of public acceptance. The armed services are beholden to a civilian Congress for their budgetary appropriations, manpower authorizations, and declaration of war. Although some personnel are careerists, the military is primarily a cadre of civilians whose commander in chief is also a civilian. As a consequence, public opinion has an influence on the military. The military is a traditional institution and typically lags behind society with regard to social change. Women and gays have gained wide acceptance in recent years as police officers and fire fighters (see chapter 7), as ministers, and in other occupations formerly limited to men.

Lack of public acceptance was addressed in the report of the Presidential Commission (1992) by five commissioners who did not subscribe to the majority opinion, as follows: "The question is whether the American people would view it as right and proper to assign women to combat arms and whether they would support the decision of the President to do so. Notwithstanding the significant strides women have made in the civilian world, assigning women to combat would be taking egalitarianism to an unprecedented extreme" (p. 59).

The other side of this issue is whether our enemies would perceive a country that uses women in combat as desperately vulnerable. Allan Carlson, in his testimony before the Presidential Commission, stated, "No significant human society since the dawn of the historical record has ever, with any degree of success, intentionally used organized female soldiers in real extended combat . . . The possible few exceptions have been among peoples under great military or political stress" (quoted in Presidential Commission's Report to the President, 1992, p. 61). Carlson's comments are not applicable to gay men.

Public opinion polls conducted during 1992 revealed an endorsement rate ranging from 40 percent to 60 percent for homosexuals serving in the military (National Defense Research Institute, 1993). Support for including homosexuals in a draft is much higher at 78 percent, suggesting that many members of the public are not convinced that military effectiveness would suffer if gays were admitted. One of the arguments against use of military women in combat was that the American people would not tolerate it if they came home in body bags. There was no public outcry, however, when several women lost their lives in the Persian Gulf War. We believe that the public would display a similar laissez-faire attitude if the ban on homosexuals were removed.

Security. This is an issue when women become prisoners of war. It has been argued that women would collaborate with the enemy in exchange for favors. Others fear that women would be raped and less able to withstand torture than men and, as a consequence, would readily reveal classified information. Some members of the Presidential Commission, thinking that women could be used as pawns by their captors to force their male peers to talk, requested testimony from men concerning women's vulnerability to such pressure.

The security argument with respect to gay men involves the threat of blackmail. It is believed that gay men would betray their country rather than allow their sexual orientation to be revealed to the military. Ironically, if gay men were allowed to serve openly in the military, they could not be blackmailed. Even under current regulations, the fear is unfounded. After reviewing more than one hundred books and journal articles, Sarbin (1991) concluded that "sexual orientation is unrelated to moral character. Both patriots and traitors are drawn from the class *American citizen* and not specifically from the class *heterosexual* or the class *homosexual*" (p. 32).

Use of the threat to national or military security as an argument against homosexuals has no basis in fact. Thousands of gays and lesbians have served loyally in the military, many as officers who retired with honor.

Lessons to Be Learned from the Integration of Women

Although women have not yet achieved the status of men in the military, a major shift toward greater equality has occurred. Three forces have brought about this change—military need, social-political climate, and defining incidents. Whether these forces might similarly promote the status of homosexuals is open to conjecture, but precedents have been established that are worth considering in regard to the integration of homosexuals.

Military Need

Shilts (1993) has described the inconsistent treatment of homosexuals in times of peace and war, demonstrating that the military overlooks its aversion to homosexuals in time of need. Since the dismemberment of the Soviet Union, scenarios of future wars have envisioned limited conflicts that would not require a massive mobilization of civilians. Thus, a World War II–like crisis, in which more than 200,000 women served, does not appear a likely scenario for allowing open homosexuals to serve in the military.

The military always needs new recruits. Because enlistments of young men have been falling since 1993, restrictions on the number of women were abolished to meet recruiting needs (Kreisher, 1995). The backlog of women who were waiting to enter the military ensured that recruiting quotas were met. Once the backlog is exhausted, however, it is likely that standards will have to be lowered to permit the entry of marginal men, or other personnel pools will have to be considered. Homosexuals do not represent a large manpower pool, but because of severe restrictions on their participation in the military, they are an untapped resource.

The problem with relying on need for access to the military is the difficulty in maintaining acceptance. With women, when the need evaporated, so did access to the military in meaningful numbers. Homosexuals, too, have experienced the revolving door of expendability. Thus, once status is achieved, it must be codified to ensure permanency.

Social-Political Climate

As pointed out earlier, the social climate in the United States is far more accepting of equal employment opportunity for homosexuals than is the military. A similar statement can be made about opportunity for women. The political climate, however, has the potential for a greater lasting influence on change than the social climate. Actions from the executive, legislative, or judicial branch of government could give homosexuals access to the military that would be difficult to rescind.

The presidential election of 1992 was precedent-setting in that the Democratic candidate, Bill Clinton, openly supported the rights of homosexuals in the military. Presidents have issued executive orders that modified the status of women and African Americans in the military, and for a time after his election it appeared that Clinton might play a similar role with regard to homosexuals. But the midterm elections of 1994 weakened the Democratic mandate, so the commander in chief is not likely to act on this controversial issue.

Congress has modified the status of women in the military by legislating and voiding restrictive statutes, and challenging the uniformed military on regulations applying solely to women. Congress also codified the "Don't Ask, Don't Tell" policy on homosexuality. The current Republican-controlled Congress, with its emphasis on family values, is not likely to champion equality for homosexuals.

The most likely branch of government to promote access to the military for homosexuals is the judicial branch. The judiciary is not quickly influenced by changes in the political climate because federal judges are appointed and typically serve until they wish to retire. The courts played a very significant role in securing greater equality for women in the military. To date, much of the progress made by homosexuals has come through the judicial process. Several celebrated cases have resulted in individual gay men and lesbians being reinstated in the military after being discharged for homosexuality. A successful class action suit could end discrimination based solely on sexual orientation, just as the 1976 class action suit ended prohibitions against Navy women serving aboard ships, based solely on gender.

Defining Incidents

The integration of women was helped by several critical events. The Persian Gulf War greatly increased public awareness of the expanded role of active-duty and reserve women. Women came under fire, died, and were taken prisoner. Yet an informed American public tolerated the risk to women, contrary to the predictions of those who argued against women in combat. The bill to permit Air Force and Navy women to fly combatant aircraft subsequently passed the House and Senate easily when the war ended.

Another defining incident was the 1991 Tailhook Convention. Again, television and print media played a major role. The widespread publicity given the flagrant behavior of some Naval aviators at the convention aroused a sense of guilt in the military hierarchy and hastened the removal of the final barriers against women in combatant ships and aircraft.

A scenario that would constitute a defining incident for gays is difficult to imagine. The 1992 hate murder of a gay sailor, seaman A. R. Schindler, Jr., by a shipmate should have been sufficiently egregious to raise an outcry. Yet it does not appear to have led to a softening of the military's stance on homosexuals.

Application to Homosexuals

Homosexuals, like military women, must be proactive if change is to occur. The three forces just discussed may appear to be externally controlled, but each creates a window of opportunity. Military women have used the political process and public opinion to bring pressure on members of Congress for changes in restrictive laws. They have filed suit against the government, appeared in uniform on Capitol Hill to influence congressional hearings, and contacted their local news media when the chain of command failed to respond adequately to injustices. Such activities are not without risk, certainly more so for closeted homosexuals than for women. But it is unlikely that an institution as traditional and conservative as the military will voluntarily admit homosexuals in its ranks.

Progress will be uneven, and acceptance may prove elusive. Women made significant gains in the 1970s, only to experience a backlash in the 1980s and a resurgence of gains in the 1990s. Yet the struggle is not over. Pregnancy among women in combatant ships has called into question the policy permitting such assignments and threatened newfound opportunities. Lesbians and gay men, too, recently have experienced victories in suits against the military, and then watched the political climate shift as conservatives gained control of Congress.

Conclusions

Women in the military and homosexuals in general are viewed as having adopted inappropriate gender roles, and as a consequence, both are members of out-groups. Neither women nor gay men have been afforded equal opportunity within the armed forces. At times the exclusionary attitudes of mainstream men defy reason, as when an Air Force chief of staff admitted that he would send a less-skilled male pilot into battle rather than a more qualified female pilot, even though combat readiness would be adversely affected (Gellman, 1991); or when a Navy vice admiral urged the rooting out of lesbians, although they were generally "hard-working, career-oriented, willing to put in long hours on the job and among the command's top performers" ("Foes of Navy's," 1990, p. A2).

Most of the barriers to equality for military women have toppled. In the early stages of women's fight for equality, military need was the primary impetus for change. Later, women became empowered and turned to the courts for redress of disparate laws and military regulations. More recently, the media and the Congress have had a profound influence.

The gay movement lags behind the women's movement in both the military and society. After the election of a Democratic president in 1992, gay and lesbian servicemembers turned to the courts with mixed success to fight their separation from the military and test the limits of the "Don't Ask, Don't Tell" policy. Moreover, homosexuals in uniform are beginning to learn how to use the media to demonstrate their numbers, accomplishments, and patriotism.

It is hoped that a critical incident such as Tailhook will not be the impetus to hasten acceptance of homosexuals in the military. Nonetheless, change is occurring. If the progress of women in the military is a valid model, sexual orientation, like sexual membership, will no longer limit participation in the defense of our nation.

References

Benecke, M. M., & Dodge, K. S. (1990). Military women in nontraditional job fields: Casualties of the armed forces war on homosexuals. *Harvard Women's Law Journal, 13,* 215–250.

Binkin, M., & Bach, S. J. (1977). *Women and the military.* Washington, DC: Brookings Institution.

Borack, J. I. (1978). *Intentions of women (18–25 years old) to join the military: Results of a national survey* (NPRDC TR-78-34). San Diego: Navy Personnel Research and Development Center.

Cropsey, S. (1980, fall). The military manpower crisis: Women in combat. *Public Interest,* 58–73.

Department of Defense. (1982). *Enlisted administrative separations.* (Directive 1332.14). Washington, DC: Government Printing Office.

DePauw, L. G. (1988). Gender as stigma: Probing some sensitive issues. *Minerva, VI* (1), 29–43.

Foes of Navy's expulsion of lesbians claim hypocrisy. (1990, September 2). *San Diego Union-Tribune*, p. A2.

Gellman, B. (1991, June 19). Officials avoid stance on combat role. *Washington Post*, p. A15.

General Accounting Office (1992). *DOD's policy on homosexuality* (GAO/NSIAD-92-98). Washington, DC: Author.

General Accounting Office (1994). *DOD service academies: More actions needed to eliminate sexual harassment* (GAO/NSIAD-94-6). Washington, DC: Author.

Gilder, G. (1986). *Men and marriage*. Gretna, LA: Pelican Books.

Gutek, B. A., & Koss, M. P. (1993). Changed women and changed organizations: Consequences of and coping with sexual harassment. *Journal of Vocational Behavior, 42,* 28–48.

Holm, J. (1982). *Women in the military: An unfinished revolution*. Novato, CA: Presidio.

Karst, K. L. (1991). The pursuit of manhood and the desegregation of the armed forces. *UCLA Law Review, 38,* 499–581.

Kreisher, O. (1995, March 4). Policy on pregnant sailors under close watch, CNO says. *San Diego Union-Tribune*, p. A21.

Letter to the editor. (1994, November 29). *San Diego Union-Tribune*, p. B7.

Martindale, M. (1991). Sexual harassment in the military: 1988. *Sociological Practice Review, 2,* 200–216.

Meyers, M. (1994). Defining homosexuality: News coverage of the "repeal the ban" controversy. *Discourse & Society, 5,* 321–344.

Mitchell, B. (1989). *Weak link: The feminization of the American military*. Washington, DC: Regnery Gateway.

National Defense Research Institute. (1993). *Sexual orientation and U.S. military personnel policy: Options and assessment*. Santa Monica, CA: RAND.

Presidential Commission on the Assignment of Women in the Armed Forces (1992). *Report to the President*. Washington, DC: Author.

Reed, F. (1994, November 21). What men really say about women. *Navy Times*, 70.

Roper Organization, Inc. (1992). *Attitudes regarding the assignment of women in the armed forces*. New York: Author.

Roush, P. (1990). Combat exclusion: Military necessity or another name for bigotry? *Minerva, VIII* (3), 1–15.

Rustad, M. L. (1982). *Women in khaki: The American enlisted woman*. New York: Praeger.

Sarbin, T. R. (1991). *Homosexuality and personnel security*. Monterey, CA: Defense Personnel Security Research and Education Center.

Shilts, R. (1993). *Conduct unbecoming: Gays and lesbians in the U.S. military*. New York: St. Martin's Press.

Stiehm, J. H. (1989). *Arms and the enlisted woman*. Philadelphia: Temple University Press.

Terpstra, D. E., & Baker, D. D. (1991). Sexual harassment at work: The psychosocial issues. In M. J. Davidson & J. Ernshaw (Eds.), *Vulnerable workers: Psychosocial and legal issues* (pp. 179–201). New York: Wiley.

Thomas, P. J. (1981). Role of women in the military: Australia, Canada, the United Kingdom, and the United States. *Catalog of Selected Documents in Psychology, 11,* 32.

Thomas, P. J. (1994, December 9). *Sexual harassment in the Navy: Trends over the past five years.* Briefing for the Secretary of the Navy's Standing Committee on Military and Civilian Women in the Department of the Navy, Washington, DC.

Webb, J. H., Jr. (1979, November). Women can't fight. *The Washingtonian,* 144–146, 273–278, 281–282.

Wright, M. (1982, June 20). The Marine Corps faces the future. *New York Times Magazine,* 74.

FIVE

Applying Lessons Learned from Minority Integration in the Military

Michael R. Kauth and Dan Landis

Arguments opposing openly gay and lesbian military personnel have shifted considerably over time, in keeping with dramatic societal and world change. But the vitriolic language of discrimination against openly gay military personnel bears a striking resemblance to the fervent opposition that was expressed in the past to racial and gender integration (Shilts, 1993, pp. 187-189). This chapter reviews the history of integration of African Americans in the military, examines the similarities across arguments against integration of African Americans and, from past integration attempts, identifies a strategy for effectively incorporating gay and lesbian service personnel.

A Brief History of Racial Integration in the Military

Preintegration

During the Revolutionary War, military personnel needs demanded that many colonies recruit African Americans, despite the fear that armed African American soldiers might turn on whites, might create disharmony among the troops, and might drive away more valued white men (Foner, 1974). Later, during the Civil War, African Americans were prohibited from military service until too few white recruits and too many white

The opinions in this chapter are those of the authors and do not represent official positions of the Department of Veterans Affairs, the United States government, or their agencies.

casualties made enlisting African Americans more tolerable to congressional leaders and Union citizens (Binkin et al., 1972). Despite their late start, about 390,000 African Americans served in the Civil War, and 38,000 lost their lives, a rate 40 percent higher than the white casualty rate (Franklin, 1980). After the Civil War, when many African Americans could not be discharged, they were assigned duties that whites generally refused. Six all-African American regiments, called Buffalo Soldiers, were stationed in far western outposts to fight threatening Indian nations (Schubert, 1993). The Buffalo Soldiers were led by white officers who considered the duties anathema to a distinguished military career.

During World War I, most African Americans were assigned non-combat duties and menial jobs, such as mess orderlies (Stillman, 1968). A small number of segregated combat units served under French command, however. These all-African American units were subject to such harsh criticism and conditions that the National Association for the Advancement of Colored People (NAACP) launched an investigation of their treatment (Hope, 1979).

Personnel needs again became a priority when United States entry into World War II appeared likely. Another concern at the beginning of the war was the costliness and inefficiency of maintaining a segregated military (National Defense Research Institute, 1993). In 1942, the Navy and Marine Corps began accepting African American recruits but assigned them to shore positions at ammunition depots and ports (MacGregor, 1981). Yet different duties for African Americans and whites at a time of war only exacerbated racial tensions. Limited integration of supply ships began in late 1944 and produced so few problems that in 1946 the Navy eliminated racial restrictions for all general-service positions. By 1948 the Air Force also realized the inefficiency of maintaining combat-ready segregated units, although opposition to integration among senior officers effectively stalled any action on the idea (MacGregor, 1981). In contrast, the Army leadership staunchly opposed the idea of racial integration.

By the end of World War II, more than one million African Americans served efficiently in various service branches (Foner, 1974). There were no reports of racial violence or noncooperation in combat zones, although reports circulated of tension over use of recreation facilities in rear areas (Lee, 1966). Overall, events in World War II demonstrated that African American soldiers were effective fighters and that "mixing of the races" would not compromise unit effectiveness (Stouffer et al., 1949). The Stouffer et al. study of unit effectiveness may have been the first social science study that contributed directly to a change in social policy. Against the common belief of the day—that African Americans lacked the ability to be good soldiers, airmen, sailors and marines—experience proved otherwise.

The war's end created another situation that may have contributed to pressure on the military to integrate. From a peak of 9 percent in the Army in World War II, African American representation dropped in 1949 to less than 6 percent of active-duty forces. The disproportionate discharging of African American men after World War II may have created pressure from discharged servicemen and civil rights groups to change the policy (Young, 1982; MacGregor, 1981).

In 1948, President Harry Truman issued Executive Order 9981, which officially desegregated the armed forces. But desegregation proceeded slowly for most service branches and with much resistance from the Army (MacGregor, 1981). Again, in the early 1950s, shortages of combat personnel to fight in the Korean War and the high costs of maintaining separate facilities for African Americans and whites made integration a rapid necessity for the Army (Berryman, 1988; MacGregor, 1981).

Postintegration

In 1951, the Army hired social scientists from Johns Hopkins University to study the impact of desegregation on unit effectiveness of troops deployed in Korea (Operations Research Office, 1954). The study, known as Project Clear, found that racial integration had no significant effect on task performance or unit effectiveness (Bogart, 1992). In fact, 89 percent of officers assigned to integrated units reported that the level of cooperation among these units was equal or superior to that of white units. Moreover, white soldiers who had greater experience with African Americans were more accepting of integration. Project Clear findings provided additional support for full integration and, by the end of the Korean conflict, the Department of Defense (DoD) had eliminated all racially segregated units and living quarters (MacGregor, 1981).

Racial discrimination was not abolished by edict, however. Prejudice and traditional antiracial practices continued even though African Americans and whites were now living and working together. In 1963, the DoD, compelled by the Kennedy administration, issued its first equal opportunity directive, which acknowledged that racial discrimination was harmful to morale and unit effectiveness (National Defense Research Institute, 1993). This directive took the unusual and very controversial step of applying sanctions to civilian organizations that discriminated against African American personnel off base (MacGregor, 1981). The directive also created civil rights offices in the DoD and service branches to monitor and implement equitable treatment of minorities. Unfortunately, these offices lacked the personnel and financial resources to perform their function, and compliance with equal treatment directives by commanders and civilians was largely voluntary (MacGregor, 1981). Moreover, complaints by minorities of inequitable treatment often went unanswered during the 1960s.

Racial tensions reached their highest levels in the late 1960s. Unofficial discriminatory practices maintained racial inequities in the military; civil rights offices were ineffectual; African Americans and whites were increasingly polarized, reflecting American society; and African Americans were becoming more vocal and violent in demanding equal treatment (Foner, 1974; MacGregor, 1981). Racial disparities in who was likely to be drafted and who served in combat units in Vietnam increased tensions. Between 1961 and 1966, 25 percent of casualties in Vietnam were African American servicemen, comparable to their representation in the military (Berryman, 1988). But African Americans represented only 11 percent of the general population aged nineteen to twenty-one years. In 1965, about one-third of all African Americans in service were assigned to combat units, and by 1967, 27 percent of all Army fatalities were African American soldiers (Day, 1983; Landis, Hope, and Day, 1984). Clearly, African Americans were concentrated in the lower pay grades that comprised most combat units. By the late 1960s, African Americans were more likely than whites to be drafted, serve in high-risk combat units, and be wounded or killed (Badillo and Curry, 1976).

During the late 1960s, racial conflict erupted in American society and among U.S. servicemen worldwide. "Fragging," or killing of white officers by African American soldiers in Vietnam, and violent race riots in rear areas in Vietnam and bases in the United States and Europe signaled uncontrolled racial friction. Yet racial incidents in Vietnam were confined to rear areas (National Defense Research Institute, 1993), where there was less combat, more free time, readily available alcohol, drugs, bars, clubs, prostitutes, newspapers and television. Anecdotal reports of white and African American combat soldiers who spent long stretches of time in the jungle on hazardous missions suggested that the common goals of survival and fighting the enemy negated racial differences that fellow soldiers in the rear found intolerable (cf. Binkin et al., 1972; Foner, 1974; Nalty, 1986).

Violent racial events in Vietnam and on military bases in the United

States and Europe prompted Secretary of Defense Melvin Laird to establish the Defense Race Relations Institute (DRRI) in 1971 at Patrick Air Force Base, Cocoa Beach, Florida. Later renamed the Defense Equal Opportunity Management Institute (DEOMI), the agency implemented an aggressive six-week race relations training program whose goals included providing knowledge of ethnic minority history and minorities' contribution to the military; exposure to psychological, social, and cultural factors related to race relations and social dynamics; and development of teaching and group skills to lead future problem-solving discussion groups (Day, 1983). The program produced such a radical change of attitudes in its first group of trainees that white graduates were called "militant," and senior officers accused the institute of "brainwashing" recruits (Day, 1983; Hope, 1979). An investigation found DRRI/DEOMI's confrontational and experiential techniques to be overzealous, and later programs were modified to focus less on attitude change and more on awareness of personal prejudicial attitudes and compliance with equal opportunity directives. Early program graduates in the mid- to late seventies reported that DEOMI training was a positive experience and suggested the need for more reality-based conflict resolution skills, refresher courses (Finman, 1978), and greater support of programs by command (Edmonds, Nordlie, and Thomas, 1978).

DEOMI's equal opportunity training expanded to sixteen weeks in the 1980s and added material on sexism, anti-Semitism, cross-cultural differences, and personnel management. From 1971 to late 1992, DEOMI trained 12,352 recruits in race-relations and equal opportunity issues (Landis, Dansby, and Faley, 1993). More than 10,000 of these were men. More than one-third came from the Army, 14 percent were from the Navy, 13 percent were from the Air Force, 3 percent were from the Coast Guard, and 1 percent were from the marines. DEOMI also developed equal opportunity training courses for government agencies outside the DoD. But only since 1987 has DEOMI engaged in systematic research to evaluate the equal opportunity climate and effects of training. A measure of equal opportunity climate, the Military Equal Opportunity Climate Survey (Landis, Dansby, and Faley, 1993), is now in wide usage throughout the services (Landis, Dansby, and Tallavigo, 1996).

Impact of Race Relations and Equal Opportunity Training

Evidence of significant improvement in equal opportunity practices was difficult to discern, and perceptions of inequitable treatment remained

high. A recent convenience sample survey (Landis, 1990) of equal opportunity climate on six military sites ($n = 1650$) found that more than 50
percent of African American men and 60 percent of white women perceived significant discrimination in military life. There were no differences
across branches. African Americans and women (19 percent and 23 percent, respectively) also judged harassment or sex discrimination to be
more likely to occur than did whites and men (12 percent and 10 percent,
respectively). Almost 20 percent of whites perceived "reverse discrimination" compared to 7 percent of African Americans. White officers consistently perceived the least amount of racial or sex discrimination of any
group. Overall, African American service personnel still perceived about
the same amount of discrimination in 1989 as in the mid-1970s, and
whites still perceived a lack of discrimination. These findings were consistent with reports from other researchers (Rosenfeld et al., 1991).

But the Landis survey (1990) also found that African Americans and
whites viewed racial segregation as less desirable than they had more than
a decade ago, and African Americans described the military's equal opportunity efforts as "good" to "very good." Whereas whites and officers
(78 percent and 71 percent, respectively) perceived higher work group
effectiveness than African Americans and enlisted personnel (62 percent
and 62 percent, respectively), these percentages were high overall, and
all groups reported similar satisfaction with career progress.

Although minorities may be encouraged by the military's equal opportunity training efforts, it is uncertain whether such programs have produced any real change in racial attitudes or employment practices. Without controlled experimental studies and systematic follow-up, it is
difficult to determine how the equal opportunity climate has been affected
by DEOMI training programs. From 1971 to 1990, African Americans
and women continued to take longer than white men to reach higher
ranks (St. Pierre, 1991). Clearly, the percentage of African American enlisted men to officers has remained disproportionate. In 1994, the proportion of African American enlisted personnel and officers in the Army was
31 percent and 9 percent (DEOMI, 1994). Furthermore, the proportion
of African American officers dropped sharply between the O-4 (Major)
and O-5 (Lieutenant Colonel) ranks, suggesting a glass ceiling. No similar
drop was evident in noncommissioned officer ranks.

Nevertheless, it may require twenty to thirty years for a second lieutenant
to attain flag rank (general officer). Therefore, changes in equal opportunity
policy in 1971 would not produce a significant effect at the highest rank
until the 1990s. Indeed, the current percentage of African Americans of flag

rank in the Air Force (1.7 percent) is almost precisely the same as the number of African Americans who were at the junior officer level in 1971 (DEOMI, 1994). Without institutional efforts to eliminate discriminatory practices that were begun in the early 1970s, the present level of accomplishment for minorities would likely be many years in the future.

One may argue that changes in the military's racial climate were inevitable. Given the violent racial conflict evident on military bases in the late 1960s and early 1970s, however, it is unlikely that announcement of a new antidiscrimination policy in 1971 would have eased tensions at all. Announcing that racial discrimination is no longer an acceptable practice does nothing to convince personnel of the policy's merits, solicit support for the policy, or eliminate *unofficial* discriminatory practices. Consistency between policy and practice is unlikely to occur spontaneously. DEOMI programming was an attempt to create consistency in policy practice and promote civility among personnel, if not attitude change.

Whether DEOMI programming has produced actual changes in the racial climate, it is clear that increased numbers of minorities are being retained in the military (DEOMI, 1994). One reason for greater minority retention could be that people from lower socioeconomic groups are waiting out a less-promising civilian economy. Minority retention rates could also be a reflection of perceived vigor in the promulgation and enforcement of antidiscriminatory military policies. Despite perceptions of persistent discrimination, minority personnel in the Landis survey (1990) reported high work group effectiveness and career satisfaction. In a survey of DEOMI graduates, Landis, Dansby, and Faley (1994) found that the decision by nonwhites to remain in the service was heavily dependent on one's perception of the equal opportunity climate in his or her unit. Perhaps attention to equal opportunity inequities and practices by command, rather than elimination of discrimination, is important to minorities in deciding whether to remain in the armed forces. An atmosphere of greater cooperation may explain why more minorities are choosing to remain in the military, although discrimination is still perceived. Equal opportunity training may teach people how to work together more effectively, but it does not change personal prejudices.

Arguments For and Against Comparison of Race and Sexual Orientation

Few would argue that the history of discrimination against African Americans in the United States is the same as the treatment of gay men and lesbians. Discriminatory practices have been applied differently to each group, owing to differences in perceptions of group characteristics, member visibility, and socioeconomic status of group members. But these differences do not negate several interesting similarities related to the groups' experiences with the military.

Prejudicial attitudes and practices in the military regarding African Americans and gays were based on characteristics unrelated to ability to work (cf., McDaniel, 1989). Rather, dark skin color and same-sex object of romantic attraction were what whites and heterosexuals, respectively, found intolerable. "Reasons" were varied for white social distance from African Americans. Whites feared that "mixing of the races" would result in an epidemic of sexually transmitted disease; an increase in antiracial violence and criminal activity by African Americans; the breakdown of morale, order, and discipline, resulting in weakened national defenses; mass exit from the military by whites; and greater difficulty recruiting whites for service (Binkin et al., 1972; Money, 1993). Whites also feared that no white enlisted man would respect or follow the orders of an African American officer. As illustration of these fears, in 1948, when the issue of racial integration was hotly debated and on the verge of certainty by President Truman's executive order, Senate Armed Forces Committee Chairman Richard Russell (D-GA) declared, "[T]he mandatory intermingling of the races throughout the services will be a terrific blow to the efficiency and fighting power of the armed services. . . . It is sure to increase the number of men who will be disabled through communicable diseases. It will increase the rate of crime committed by servicemen" (Congressional Quarterly, 1972, pp. 34–35).

Arguments against racial integration in the 1940s sound curiously like pronouncements today against allowing openly gay personnel in the military. Proponents of the gay ban have argued that greater tolerance of gay men would promote the spread of human immunodeficiency virus (HIV); increase violence against gay men and lesbians; diminish morale (of heterosexuals); contribute to misconduct by heterosexuals who would refuse to follow orders from gays; give the appearance of a weakened (effeminate) military; result in heterosexuals leaving service; and create problems recruiting heterosexuals in the armed forces (Kauth and Landis, 1994). Like those who argued against racial integration, proponents of the gay ban have provided no empirical evidence for their predictions and have relied on the emotional appeal of their statements to persuade.

A further similarity between racial integration and acceptance of openly gay men and lesbians in the military is the social context surrounding these debates. Although the DoD argues that race, unlike sexual orientation, is a "neutral characteristic" (cf. Keating, 1991), this was by no means the prevailing attitude in 1948, when Truman mandated integration. In a 1948 Gallup Poll of three thousand adults, 63 percent of those surveyed favored racial segregation of the military. Only 26 percent of the sample supported integration. In the 1940s, racial integration was counter to societal norms, which held to strict social and environmental segregation of whites and African Americans. Religious organizations used a misreading of the biblical story of Cain and Abel to justify separation of whites and African Americans (Allport, 1954; Jordon, 1968). Places of business posted signs stating "Whites Only," or separate and inferior facilities were labeled "Coloreds," if they were available at all. In such an atmosphere, a southern-dominated Congress was not about to pass legislation to desegregate the armed forces. Although President Truman's executive order to integrate the military was no doubt influenced by pressure from civil rights groups (as well as his own personal and philosophical beliefs), it was politically risky in an election year. Truman was vilified for his actions, and the attacks were often accompanied by dire predictions about the weakening of the armed forces and national security (see Russell's comment above). Many may have missed the irony of a segregated U.S. military fighting a country devoted to ethnic annihilation in World War II.

The climate surrounding debate over gays in the military has been no less charged than that over racial integration. Although gay people today are not overtly segregated from society as African Americans were in the 1940s, gays benefit most from social privileges by masking as heterosexual. Since the rise of the gay rights movement in the late 1960s, gay men and lesbians have been less willing to deny or hide their identity or romantic relationship from heterosexuals. An increasingly visible gay community has demanded the same opportunities and privileges afforded to heterosexuals, and these claims have clashed in a social border war with religious fundamentalist and conservative political forces. Writing for the DoD publication *Parameters*, Major R. D. Adair and Captain Joseph Myers (1993) described the issue over gays in the military as a "cultural civil war" that "threatens to divide the nation" (p. 10). The war's present front is the military's ban on openly gay personnel, and the battle is over the sexual threat to heterosexual men by gay men. Worried military leaders have voiced publicly their positions and fired their own salvos. In 1992, General Colin Powell, chairman of the Joint Chiefs of Staff, stated,

"It's difficult in a military setting where there is no *privacy* . . . to introduce a group of individuals—proud, brave, loyal, good Americans, but
who favor a homosexual lifestyle—and put them in with heterosexuals
who would prefer not to have somebody of the same sex find them sexually attractive" ("Powell Defends," 1992; emphasis added).

Similarly, the 1993 congressional hearings on gays in the military,
chaired by Senator Sam Nunn (D-GA), presented few empirical findings
but elicited strong emotions from the panel and television viewers by emphasizing the physical intimacy servicemen share. The flash point of the
hearings may have occurred when Nunn led an entourage of congressional panel members and television crews on a tour of a cramped submarine. A photograph of the tour, which was published in newspapers
all over the country, featured the submarine's close sleeping quarters,
suggesting that heterosexuals would have to sleep side by side with gays.

The Butler Argument

Despite several similarities, many African Americans have resisted comparing racial discrimination to the experience of gays (Williams, 1993).
Therefore, it may be helpful to examine an argument against the comparison. One prominent opponent of the racial analogy is academician and
sociologist John Butler (1993), in a recent issue of *Society* (see also Adair
and Myers, 1993; Moskos, 1993; Keating, 1991).

Butler stated that the context of integration differed for African Americans. Despite major gains for African Americans in establishing status in
the military, racial integration was undertaken as a result of "manpower"
rather than social needs (p. 14). He noted that the military was not "set
up to solve issues of social behavior" (p. 15). In addition, Butler claimed,
people who compare the African American experience to oppression of
gay men and lesbians in the military "usually do so for political reasons,"
implying that politics had little to do with racial integration of the military in 1948 (p. 17).

Second, Butler declared that discrimination against African Americans
was fundamentally different from prejudice against gays. The racial metaphor cannot be applied to gays because "men and women who engage
in homosexual behavior do not make up a separate racial or ethnic
group" (p. 14). Furthermore, "one cannot compare an *achieved behavior*
that runs through all racial groups with an *ascribed characteristic* like
race" (p. 17; emphasis added). As an achieved behavior, Butler said, homosexuality is "a condition that can be treated" (p. 18).

And finally, Butler viewed homosexuality as largely a phenomenon among whites and, therefore, charges of denied civil rights were untenable. Moreover, comparing the experiences of African Americans to gay men and lesbians "trivialize(s) the Black experience" (p. 16). In his opinion, "White homosexuals are well represented in all professions and have had the same opportunity as other white persons to take advantage of the varied existing opportunity structure" (p. 18). Being an issue among privileged whites, treatment of gay people "cannot be based on arguments about civil rights and the denial of civil rights" (p. 19).

Removed from their emotional trappings, as above, Butler's statements may be less convincing to those opposed to the racial metaphor. Rather than provide reasons for its inappropriateness, Butler simply stated that people who made the comparison did it "for political reasons" (p. 17). Worse, Butler failed to acknowledge that events in the social context of the 1940s provided pressure for the military to change its policy on segregation of African Americans. The country was vehemently opposed to integration of the military and emerging African American civil rights groups, such as the NAACP, lobbied hard with President Truman and Congress for integration. Although other factors were important in the change of policy, within the social atmosphere of the 1940s, it would be difficult to deny the political influence on Truman's decision. Butler also forgot that the change of policy occurred during peacetime, prior to buildup for the Korean conflict. Buildup of personnel for war aided integration, but it was not the event that initiated integration.

More critical to his position, the crux of Butler's argument rests in his description of homosexuality as an "achieved behavior" (p. 17). Butler's choice of terms is important to understanding his position. "Achieved" connotes "learned" or somehow under volitional control. "Behavior" suggests observable activity. For most social scientists, behavior may be observable or nonobservable (as in internal but measurable actions) or even dispositional. For the layperson, however, behavior is something that someone does and will do. The assumptions beneath Butler's thesis are that gay people are (a) gay by choice and (b) are impelled to act on their orientation no matter what the setting. Viewed in this way, Butler's argument is patently false. Even if (a) is true, (b) does not necessarily follow. Indeed, (a) can be either true or false without implying (b) at all. If Butler's primary concern is the regulation of sexual behavior, then sexual orientation is irrelevant to achieving that goal.

Finally, Butler argued that comparing the experience of gay people to that of African Americans trivialized the pain of slavery and its aftermath. But Butler confused the etiology of discrimination with its outcome. Again, not many would argue that gay people have gone through the same experience as African Americans, although this distinction might be lost on those who wore a pink triangle in Nazi concentration camps. Instead, advocates of policy change have looked at the origins of prejudice and the pain of oppression to both the individual and society, and these circumstances are what bear a haunting similarity for both groups. To an extent, Butler committed his own grievance: He trivialized the experience of gay men and lesbians, especially ethnic gay men and lesbians who often experience double prejudice because of their ethnicity and sexual orientation. In fact, Butler failed to recognize discrimination suffered by ethnic gay men and lesbians and implied that under certain circumstances prejudice could be justified. Even worse, Butler suggested that prejudice and discrimination against those out-groups that we do not like, such as gay people, was perfectly reasonable.

Dissimilarities with the African American Experience

There are also several differences between the experiences of African Americans and gays in the military that deserve note. For example, proponents of excluding gay men and lesbians from military service have argued that their presence could threaten national security if closeted gays were blackmailed with exposure by foreign powers (although this argument has been abandoned recently), lead to fraternization and sexual harassment of young heterosexuals, and violate the sexual privacy of heterosexuals (Shilts, 1993). Again, proponents of the antigay policy have provided no more empirical support for these arguments than they have for others. Significantly, concerns about fraternization, harassment, and privacy are similar to arguments used to prohibit gender integration (Berryman, 1988; Marsden, 1986; Stiehm, 1989; see also chapters 4 and 11). Another interesting aspect of objections to tolerance of openly gay people is the (presumed) target of threat (by sexual attraction and advances): heterosexual men (see quote by Powell above). While Nunn was pointing out the intimate sleeping quarters available on a submarine to camera crews, Norfolk Naval base Commander Lin Hutton commented that if the gay ban were dropped "[m]any people would worry that they might be the target of unwanted sexual advances, actual or perceived" ("Nunn offers," 1993, p. 1241). Sexual conduct and perceived advances by parties of any sexual orientation will be important issues to address if a policy more tolerant of gays is effectively implemented.

Another distinction between African American and gay experiences is the degree of public opposition to full integration. Public opinion polls show Americans to be less opposed to gays in the military today than they were to racial integration in the 1940s. As mentioned earlier in this chapter, a 1948 Gallup poll reported that 63 percent of those surveyed favored racial segregation in the military, whereas only 26 percent supported integration. But recent polls show Americans to be divided on the issue of gays in the military and responses depend largely on how the questions are asked. For example, a January 1993 Gallup poll of 774 adults found that 53 percent answered positively to the statement, "Should homosexuals be able to serve in the Armed Forces?" Yet a July 1993 Gallup poll of 1,002 adults revealed that only 48 percent of those surveyed disagreed with the statement, "Homosexuality is incompatible with military service." At the same time, 49 percent opposed the "Don't Ask, Don't Tell" policy of discharging service personnel who disclosed their homosexuality. In addition, a *Wall Street Journal*/NBC News poll (June 1993) of 1,502 registered voters found that 79 percent (after poll rounding) supported allowing gay people to serve in the military: 38 percent supported gay people serving only if they kept their orientation private, and 40 percent agreed that gay people could serve openly. Even religious organizations are split. The Evangelical Lutheran Church favored dropping the military ban and cited its policy of ordaining priests based on their behavior rather than sexual orientation (Chilstrom, 1993). The Southern Baptist Christian Life Commission, however, strongly opposed lifting the ban because it "will give approval and support to an immoral, harmful lifestyle" ("Baptists Call," 1993). A nation evenly divided on whether to lift the ban should not be mistaken for one holding opinions that are less strong on a less controversial issue. Moreover, implementation of a policy more tolerant of gay service personnel may be no less difficult than racial integration.

Lessons Learned from Racial Integration and their Application to Integration of Openly Gay Service Personnel

Several lessons can be drawn from the experience of racial integration of the armed forces, which may assist implementation of a more tolerant policy toward openly gay personnel:

1. Segregation of African Americans was based on factors (skin color and the discomfort of whites) that were not directly related to the capacity of African Americans to perform their job. We would argue that the exclusion of gay men and lesbians is also based on factors unrelated to job performance (McDaniel, 1989).

2. Integration of a social out-group can occur in a hostile military environment and without public support. Although some branches were more supportive of integration than others and integration had made some gains before Truman's order in 1948, support from military leaders was not uniform and the opposition was strong. But military commanders in support of integration were essential to creating an atmosphere of equitable treatment. It is also noteworthy that the military was integrated before the country experienced desegregation. Moreover, public opinion polls show the country was more opposed to racial integration than it is to gays in the military.

3. A powerful incentive for racial integration in the military was the costliness and inefficiency of maintaining segregated units. Although personnel demands during times of combat facilitated movement toward integration, peacetime often brought renewed social restrictions. It seems unlikely that integration would have occurred on its civil rights merits alone. Likewise, advocates of a gay-tolerant policy may win more supporters by focusing on the economic and personnel costs of excluding gay men and lesbians, rather than on an issue of civil rights.

4. Rapid integration of a social out-group during the Korean War resulted in few problems, perhaps because of the shared goal of fighting a common enemy. (For the consequences of slow social integration during peacetime, see discussion on gender integration in chapter 4.)

5. Despite predictions about the loss of morale, order and discipline, racial integration did not reduce the effectiveness of the armed forces. Even in a time of great racial strife such as the late 1960s, combat soldiers in Vietnam who shared a goal of survival could work together effectively despite ethnic differences. Common task-related or superordinate goals, personal contact, and equal status may be more important to unit effectiveness than racial or sexual orientation similarity among group members (Allport, 1954; Amir, 1976; National Defense Research Institute, 1993). That gay people have been and are in the military at various levels could facilitate tolerance of gays, if the environment supports tolerance as a goal.

6. The military institution is resilient. It has adapted to dramatic reor-

ganization in the past fifty years, including racial integration, gender integration, and a switch to an all-volunteer force. In light of these changes, it seems unlikely that the military could not adapt to openly gay personnel among its ranks. Indeed, gay people have served and continue to serve honorably in the military—some serving secretly and others not so secretly (see Shilts, 1993).

7. Policy change does not eliminate prejudicial attitudes or practices. Equal opportunity training may facilitate civility and good working relationships, but may have little effect on personal prejudices. Even so, equal treatment will not immediately result in greater minority representation at high levels. Like African Americans, openly gay recruits will require several years to advance through the ranks.

8. The perception of efforts toward improvement of equal opportunity conditions may be more important to retention of minorities in the military than elimination of discrimination. DEOMI programming may have played a vital role in creating the perception among minorities of a more tolerant environment. It is unclear at present what components of training are most effective in accomplishing this perception. But one important factor in creating an impression of equal treatment is the monitoring and enforcement of antidiscriminatory practices. In part, racial tensions were exacerbated in the 1960s by policies that were not enforced.

9. Initially, equal opportunity training by DEOMI took a bottom-up approach. Coincidentally, those (white officers) at the top who were charged with implementing policy may have been less sensitized to discriminatory practices than African Americans and enlisted personnel. Recognizing that top-down interventions targeting those in command may be necessary, Secretary of Defense William Perry ordered DEOMI in 1994 to train all newly selected admirals, general officers, and senior executive personnel.

10. National leadership is crucial to effective implementation of outgroup integration. Once President Truman had formulated a strong position on racial integration, the tradition of military compliance to civilian authority led to efforts to produce change. A movement toward racial integration in the military was growing before Truman's action, however, and commanders who favored integration played a vital role in implementing the new policy. Implementing a policy tolerant to gays will likely require strong civilian authority from the president or the Supreme Court, as well as strong military leadership.

11. Religious opposition to social integration was deflected by a focus

on civility rather than changing personal beliefs. Traditionally, the military has drawn many of its members from areas in the country where conservative denominations are most prevalent. These same regions were most opposed to racial integration of the military fifty years ago. A similar focus on working relationships rather than acceptance of gays may be similarly effective in handling religious opposition.

Where the Racial Analogy Fails

As we have already pointed out, the comparison of African American and gay experiences is not perfect. Although noting similarities in the histories of African Americans and gays may prevent us from repeating previous mistakes, differences in their experiences will direct us to new ground. One distinct difference between the groups is that African Americans are visible. The characteristic used to distinguish them—skin color—was public, although this criterion was not always so obvious. Because everyone could know who was "black," it was easier to keep blacks together and away from whites. The overt maneuvering and energy expended in separating people by skin color may also have made it easier to recognize racial discrimination as a problem and attempt to resolve it.

Gay people, however, are not as identifiable as African Americans, which in some ways make gays more threatening to heterosexuals. Stereotypes of gay appearance and behavior are limited in utility. It is much harder to talk about gay people and their mistreatment when we do not recognize them.

A second difference between African Americans and gays is that African Americans have largely occupied the lower socioeconomic classes and have borne the stigma associated with being poor. Thus, racism in the United States was linked with social class, and class polarization may have aided recognition of biased social practices. But gay people do not predominate a particular social class and may be found across all social strata, including ethnicity. It is much more difficult to argue that a "class" is disadvantaged when not all people are disadvantaged.

The last distinction is related to the second. Whereas skin color is a dominant genetic characteristic and family members generally look similar, homosexuality, if hereditary, is not dominant because most family members are heterosexual. Most African Americans have African American parents, African American family members, and an African American social community; most people in that community will be recognized as African American. Shared lives (and shared oppression) can be a source of strength, as African American communities have demonstrated. But gay men and lesbians are not born into a community and must find each other to experience the strength and safety of community. Being isolated as a gay person without support from a community may increase vulnerability to discrimination.

Summary

From the differences between the experiences of African Americans and gays, advocates of military policy change can know where they need to focus further attention. From the similarities between African Americans and gays, the history of racial integration in the military can provide several lessons that could facilitate implementation of a more gay-tolerant policy. Yet lessons from integration should not be taken to mean that acceptance of openly gay personnel in the military will be easy, painless, or peaceful.

References

Adair, R. D., & Myers, J. C. (1993, Spring). Admission of gays to the military: A singularly intolerant act. *Parameters*, 23, 10–19.

Allport, G. (1954). *The nature of prejudice*. New York: Addison-Wesley.

Amir, Y. (1976). The role of intergroup contact in change of prejudice and ethnic relations. In P. A. Katz (Ed.), *Towards the elimination of racism* (pp. 245–308). New York: Pergamon.

Badillo, G., & Curry, G. D. (1976). The social incidence of Vietnam casualties: Social class or race? *Applied Forces & Society*, 2, 397–406.

Baptists call for keeping military ban on gays. (1993, June 5). *The Los Angeles Times*, p. B4.

Berryman, S. E. (1988). *Who serves? The persistent myth of the underclass army*. Boulder, CO: Westview Press.

Binkin, M., Eitelberg, M., Schexnider, A. J., & Smith, M. M. (1972). *Blacks and the military*. Washington, DC: Brookings Institution.

Bogart, L. (Ed.). (1992). *Project Clear: Social research and the desegregation of the United States Army*. New Brunswick, NJ: Transaction Books.

Butler, J. S. (1993, November/December). Homosexuals and the military establishment. *Society*, 31, 13–21.

Chilstrom, H. (1993, February 2). *Letter to President Clinton*. (Cited in National Defense Research Institute, 1993).

Congressional Quarterly. (1972). *The power of the Pentagon.* Washington, DC: Author.

Day, H. R. (1983). Race relations training in the U.S. military. In D. Landis and R. Brislin (Eds.), *Handbook of intercultural training: Volume II* (pp. 241–289). Elmsford, NY: Pergamon.

Defense Equal Opportunity Management Institute. (1994). *September 1993 semi-annual race/ethnic/gender profile of the Department of Defense Active Forces, Reserve Forces.* (Statistical Series Pamphlet No. 94-2). Directorate of Research, Defense Equal Opportunity Management Institute, Patrick Air Force Base, Cocoa Beach, FL.

Edmonds, W. S., Nordie, P. G., & Thomas, J. A. (1978). *Analysis of individual race relations and equal opportunity training in army schools.* (ARI Technical Report TR-78-B15). Alexandria, VA: U.S. Army Research Institute for the Behavioral and Social Sciences.

Finman, B. G. (1978). *An analysis of the training of army personnel at the Defense Race Relations Institute.* (ARI Technical Report TR-78-B14). Alexandria, VA: U.S. Army Research Institute for the Behavioral and Social Sciences.

Foner, J. D. (1974). *Blacks and the military in American history.* New York: Praeger.

Franklin, J. H. (1980). *From slavery to freedom: A history of Negro Americans* (5th ed.). New York: Knopf.

Gallup Organization. (1948), May 28-June 2). *Survey of 3,000 adults based on personal interview.* (Available from The Roper Center for Public Opinion Research, University of Connecticut, Storrs, CT.)

Gallup Organization. (1993, July 9-11). Support compromise plan? Survey of 1,002 adults. *Gallup Poll Monthly,* 30–31.

Gallup Organization/*Newsweek.* (1993, January 28-29). *Survey of 774 adults based on representative telephone poll.* (Available from The Roper Center for Public Opinion Research, University of Connecticut, Storrs, CT.)

Hope, R. O. (1979). *Racial strife in the U.S. military.* New York: Praeger.

Jordon, W. D. (1968). *White over Black: American attitudes toward the Negro, 1550-1812.* Chapel Hill, NC: University of North Carolina Press.

Kauth, M. R., & Landis, D. (1994, July). The U.S. military's "Don't ask, Don't tell" personnel policy: Fear of the open homosexual. In D. Landis (Chair), *Prejudice and discrimination in large organizations.* Symposium conducted at the Second International Congress on Prejudice, Discrimination and Conflict, Jerusalem, Israel.

Keating, T. D. (1991, May 3). *Reply to 40 members of Congress from the Office of Assistant Secretary of Defense, Director of Legal Policy.*

Landis, D. (1990, January). *Military equal opportunity climate survey: Reliability, construct validity, and preliminary field test.* Oxford, MS: Center for Applied Research and Evaluation, University of Mississippi.

Landis, D., Dansby, M. R., & Faley, R. (1993). The military equal opportunity climate survey: An example of surveying in organizations. In P. Rosenfeld, J.E. Edwards, and M.D. Thomas, (Eds.), *Improving organizational surveys:*

New directions, methods and applications (pp. 210–239). Newbury Park, CA: Sage.

————. (1994, April). *The relationship of equal opportunity climate to military career commitment: An individual differences analysis using latent variables.* Paper presented at the Biennial Behavioral Science Symposium, U.S. Air Force Academy, Colorado Springs, CO.

Landis, D., Dansby, M. R., & Tallavigo, R. (1996). The use of equal opportunity climate in intercultural training. In D. Landis and R. Bhagat (Eds.), *Handbook of intercultural training* (2nd ed., pp. 244–263). Thousand Oaks, CA: Sage.

Landis, D., Hope, R. O., & Day, H. R. (1984). Training for desegregation in the military. In N. Miller and M. B. Brewer (Eds.), *Groups in contact: The psychology of desegregation* (pp. 257–78). Orlando, FL: Academic Press.

Lee, U. (1966). *United States Army in World War II: Special studies, employment of Negro troops.* Washington, DC: U.S. Army Office of Military History.

MacGregor, M. J., Jr. (1981). *Integration of the armed forces, 1940–1965.* Washington, DC: U.S. Army Office of Military History.

Marsden, M. A. (1986). The continuing debate: Women soldiers in the U.S. Army. In D. R. Segal and H. W. Sinaiko (Eds.), *Life in the rank and file: Enlisted men and women in the armed forces of the United States, Australia, Canada, and the United Kingdom* (pp. 58–78). Washington, DC: Pergamon-Brassey's International Defense Publishers.

McDaniel, M. (1989). *Preservice adjustment of homosexual and heterosexual military accessions: Implications for security clearance suitability.* (PERS-TR-89-004). Monterey, CA: Defense Personnel Security Research and Education Center.

Money, J. (1993, November/December). Parable, principle, and military ban. *Society, 31,* 22–23.

Moskos, C. C. (1993, Winter). From citizens' army to social laboratory. *The Wilson Quarterly, 17,* 83–94.

Nalty, B. C. (1986). *Strength for the fight.* New York: Free Press.

National Defense Research Institute (1993). *Sexual orientation and U.S. military personnel policy: Options and assessment* (pp. 368–394). Santa Monica, CA: RAND.

Nunn offers a compromise: "Don't ask, don't tell." (1993, May 15). *Congressional Quarterly Weekly Report,* pp. 1240–1242.

Operations Research Office, Johns Hopkins University (1954). *Utilization of Negro manpower.* Chevy Chase, MD: Johns Hopkins University.

Perry, W. (1994, March 3). Equal opportunity (EO). *Memorandum from the Office of the Secretary of Defense.*

Powell defends military gay ban. (1992, February 6). *Washington Post,* p. A2.

Rosenfeld, P., Thomas, M. D., Edwards, J., Thomas, P. J., & Thomas, E. D. (1991). Navy research into race, ethnicity, and gender issues: A historical review. *International Journal of Intercultural Relations, 15,* 407–426.

Schubert, F. M. (1993). *Buffalo soldiers, braves and the brass.* Shippensburg, PA: White Mane.

Shilts, R. (1993). *Conduct unbecoming: Gays and lesbians in the U.S. military.* New York: Fawcett Columbine.

Stiehm, J. H. (1989). *Arms and the enlisted woman.* Philadelphia: Temple University Press.

Stillman, R. J. (1968). *Integration of the Negro in the U.S. armed forces.* New York: Praeger.

St. Pierre, M. (1991). Accession and retention of minorities: Implications for the future. *International Journal of Intercultural Relations, 15,* 469–490.

Stouffer, S. A., Lumsdaine, A. A., Lumsdaine, M. H., & Williams, R. M., Jr. (1949). *The American soldier: Two volumes.* Princeton, NJ: Princeton University Press.

Wall Street Journal/NBC News (1993, June 5-8). *Survey of 1,502 registered voters based on representative telephone poll.*

Williams, L. (1993, June 28). Blacks reject gay rights fight as equal to theirs. *New York Times,* pp. A1, A12.

Young, W. L. (1982). *Minorities and the military.* Westport, CT: Greenwood Press.

The Experience of Foreign Militaries

Paul A. Gade, David R. Segal, and Edgar M. Johnson

With the election of Bill Clinton as president of the United States in the fall of 1992, his campaign promise to lift the ban on homosexuals serving in the military immediately became a very hot issue for the Department of Defense (DoD) and for the military services. The first reaction of the Army, in November 1992, was to examine how foreign countries were dealing with the issue of homosexuals in their military services. The Air Force reaction was to conduct a survey of service members in late 1992 to assess how they felt about lifting the ban and how they were likely to react to it. The Army and the Marine Corps also attempted to conduct surveys of their members' attitudes toward homosexuals in the military and their reactions to lifting the ban in early 1993, but terminated these efforts under DoD order.

The Clinton administration immediately sought to lift the ban as promised. As part of the execution of the administration's plan, the new leaders of the DoD directed the military services not to conduct any surveys of their members about lifting the ban, and prohibited the release of the results of the Air Force survey. The rationale was that such surveys would show only the expected result that service members were very much opposed to lifting the ban, and would serve to polarize further a

The views, opinions, and/or findings in this article are solely those of the authors and should not be construed as an official Department of the Army or Department of Defense position, policy, or decision, unless so designated by other official documentation.

very sensitive issue. Therefore, the only officially acknowledged research conducted by the military services concerning the issue of homosexuals in the military was a small project conducted by Gwyn Harries-Jenkins at Hull University for the U.S. Army Research Institute for the Behavioral and Social Sciences (ARI) to assess the experience of foreign militaries with this issue (Segal, Gade, and Johnson, 1994).

In 1993, Senator John Warner (R-VA) requested that the General Accounting Office (GAO) study the policies, practices, and experiences of foreign militaries in dealing with the service of homosexuals. In March of that year, the DoD contracted with the RAND Corporation to conduct a comprehensive review of the issues surrounding gays in military service to prepare for the Senate hearings that were to be conducted under the direction of Senator Sam Nunn (D-GA) on the issue of lifting the ban. As part of this comprehensive review, RAND conducted yet a third investigation of what foreign militaries were doing and had done about this issue.

The purpose of this chapter is to review what we know about gay men and lesbians in the military services of other Western nations and to extend that knowledge to the U.S. military services where possible. Although there are many differences between the U.S. case and those of our Western allies, there are, we believe, many social and cultural similarities that make such comparisons useful for considering the issues that the U.S. might face by including homosexuals in the military services. Building on our earlier work (Segal, Gade, and Johnson, 1994), we begin with a summary of our major findings, followed by more elaborate information about various countries that have been examined in the greatest detail.

By combining the outcomes of studies conducted by ARI, GAO, and RAND, we are able to discuss in depth the policies and practices of eleven nations that have implications for the United States. To accomplish this, we describe and summarize the nature and approach used by each of these three organizations to study the issue of homosexuals in foreign militaries. We then discuss the policies and practices of each of the eleven countries that were studied in depth. Finally, we provide conclusions and lessons learned based on all of the studies we reviewed.

What We Know

All of the studies we have examined seem to agree that regardless of national policy, some individuals with homosexual orientations have managed to serve undetected in the military forces of virtually all Western nations. There is also consensus that most homosexuals in the military do not "come out," preferring that their sexual orientation remain a private matter. Even where policy and practice allow gay men and lesbians to serve, very few serve openly as homosexuals. There is still a stigma attached to being homosexual even in the most liberal societies, and there may be career costs associated with declaring oneself to be homosexual. Practices in countries with more liberal policies are often such that homosexuals serving openly in the military may find themselves referred for psychiatric counseling and excluded from combat units or from certain assignments.

The social-historical context of the development of policies toward minorities, women, and homosexuals is key to understanding a country's policies and practices toward homosexuals in their military services. More liberal countries seem to have had a long history of struggling with homosexual issues, and the liberalization of attitudes toward homosexuals in the military has evolved slowly, usually lagging behind more general societal attitudes. Although they are usually antimilitary, gay and lesbian activist groups often have served as the catalyst for changing policies and practices in the military services of more tolerant countries.

Most countries are more conservative in practice than they are in policy. Even in countries whose policies do not permit questions about sexual orientation at accession or during military service, the practice of not asking may be more of a conscious and considered omission designed to avoid more complex issues, such as benefits for partners and adoption rights, than it is an act of accepting homosexuality. In this manner, the more difficult issues are silenced. In fact, the most common pattern seems to be that military forces do not ask about sexual orientation, even when they have exclusionary policies.

Policies and practices about homosexuals in the military seem to follow, but lag behind, national social norms. Heterosexuality is clearly the dominant norm in Western societies. Homosexuals in the militaries of even the most liberal societies most often choose not to identify themselves openly as homosexuals at all. Antihomosexual feelings run high among the majority of service members even in countries with the most liberal policies and conditions. In every case, women are more tolerant of homosexuals in military service than are men.

Perhaps because heterosexuality is the dominant norm, no country, no matter how liberal, seems to know the precise percentage of its military that is homosexual. Moreover, none of the countries that have been studied seem to know what percentage of its general population is homosexual, although estimates range from a low of 1 percent to a high of 10 percent.

To understand how we came to these conclusions, we need to turn our attention to the development of the more detailed studies and their results. We begin by examining how each of the three organizations, ARI, GAO, and RAND, conducted their research.

The Army Research Institute

In November 1992, ARI contracted with Harries-Jenkins to examine the policies, practices, and problems of selected Western nations. Countries were selected to provide a wide range of approaches to the issue from liberal to conservative.

The approach used by Harries-Jenkins was to ask noted social scientists from the selected countries who were knowledgeable about the military to prepare case studies of their countries. Each scientist prepared a paper that described the policies and practices in his or her country. These papers also provided a social-historical review of the development of these policies and practices, and described problems and solutions that had been encountered in the development and implementation of those policies and practices.

These social scientists met to discuss their papers in April 1993. In this meeting, a common framework for analysis and presentation of the materials was adopted. The scientists met again in July 1993 to present and discuss their revised reports. The papers of the original meeting have been published as an ARI report (Harries-Jenkins, 1993). The final report will also be published as an ARI report (Harries-Jenkins, in press-a).

In addition, ARI contracted with a noted Canadian military sociologist to provide a comprehensive review of the Canadian situation similar to that provided by the Harries-Jenkins group. The results of this review are also available as an ARI report (Pinch, 1994).

In all, eight nations were studied in detail. Pinch's study covered Canada. The Harries-Jenkins (1993 and in press-a) reports covered France, Italy, Belgium, Germany, Denmark, the United Kingdom, and the Netherlands.

The General Accounting Office

As part of its mission the GAO frequently conducts special studies and investigations for members of Congress. The report "Homosexuals in the military: Policies and practices of foreign countries" was just such a special report completed by the National Security and International Affairs Division of the GAO for Senator John Warner (General Accounting Office, 1993).

The GAO began its research on foreign military policies and practices by selecting a sample of twenty-nine countries that had at least fifty thousand military service members on active duty. Official information was obtained about the position, policies, laws, and regulations concerning the service of homosexuals from government officials and embassies. Four countries declined to participate for a variety of reasons, including the sensitivity of the issue. From the remaining twenty-five countries, five, Canada, Germany, Israel, Sweden, and France, were selected for a more detailed review. These countries were selected because their official policies allowed homosexuals to serve and they shared cultural and military characteristics similar to those of the United States. The French government declined to provide more detailed information and was dropped from further study.

In the four remaining countries, military personnel, veterans, social scientists from the countries involved and the United States, U.S. embassy personnel, and representatives from homosexual advocacy groups were interviewed. Only Sweden permitted active-duty military personnel to be interviewed. The social scientists interviewed were from academia, the military services, well known think tanks, and professional organizations. Two social scientists, one from Canada and one from Germany, who participated in the GAO review were key participants in the ARI studies as well. In addition, polling organizations such as Roper and Gallup were contacted for input on the issue. Greater detail on how the GAO conducted this study is available in the appendices of the GAO report (GAO, 1993).

RAND

The chapter of the RAND report (Kahan et al., 1993) concerned with cross-national comparisons of military homosexual policies and practices is somewhat more inclusive than the GAO report in that France, the Netherlands, Norway, and the United Kingdom were incorporated into the study. As in the GAO study, Canada, Germany, and Israel were also included, but not Sweden. These seven countries were chosen because they represented a range of Western culture countries whose policies ranged from permissive to exclusionary and had some unique approach or aspect of policy that RAND researchers thought was worthy of in-depth investigation.

Researchers visited each of the seven countries and conducted interviews, where possible, with uniformed military personnel, members of the ministry of defense, civilian experts, members of parliament, and representatives of homosexual groups. Many of the personnel interviewed overlapped with those participating in the ARI and GAO research projects. The French government declined to participate in the RAND study, as it had with the GAO study. But RAND researchers were able to interview a few French authorities and to gather some documentation regarding the issue of homosexuals in the French military.

Among the three organizations, ARI, GAO, and RAND, eleven nations were studied in detail. Below, we briefly examine the combined results for those eleven nations.

Patterns in Foreign Militaries

The integration of homosexuals in the military is not an issue that is unique to the United States. As in the United States, policies regarding homosexuals in the military in other countries usually reflect, but lag behind, policies toward homosexuals in their host societies. The decriminalization of homosexuality and decisions by medical and psychological associations in Western countries to stop classifying homosexual behavior as pathological have set the stage for the inclusion of homosexuals in military service. Yet psychology has neither a long research tradition nor a rich data base on the policies and practices toward homosexuals that exist cross-nationally, a base that might be used to inform the policy debate (Herek, 1993). This chapter represents one of the first efforts to create that knowledge base.

Two general patterns emerged as we began to describe the ways in which foreign countries have dealt with the issues of homosexuals in their military forces. One involves the policies and practices of the military itself. A temporal difference may arise between the policies of military accession and the conditions of subsequent service. For example, there may be no restrictions on homosexuals entering the military but heavy sanctions on behavior while in the service (see the table below, drawing on information from the eleven nations included in the ARI, RAND, and GAO reports). In addition, the conditions of subsequent service may vary, depending on whether the soldier is a conscript or a volunteer.

Policies and Practices for Homosexuals' Military Service by Country

| Country | Accessions | | In Service | |
	Policy	Practice	Policy	Practice
Canada (Volunteer)	1988: homosexuals accepted with promotion and assignment restrictions. 1992: all restrictions removed.	No restriction or questions.	No discrimination permitted. Military promoted compliance with, and acceptance of, policy and planned to address attitudes of military toward homosexuals.	No attempts to change attitudes of military toward homosexuals. No problems reported. No major "coming out."
U.K. (Volunteer)	Since June 1992 enlistees are not asked about sexual orientation. Homosexuality is considered to be incompatible with military service, however.	Recruits will be rejected if they volunteer that they are homosexual.	Sexual Offenses Act of 1967 legalized homosexuality for those over 21 but did not prevent it as a military offense. • Lesbian acts not criminal • Homosexuals not tolerated	Administrative discharge the usual way of dealing with homosexuals; disruptions not tolerated.
France (Conscription)	No discrimination. No specific policy regarding homosexuals.	Homosexuality seen as a behavioral disorder. Can declare oneself unfit because of homosexuality. No alternative service required.	No specific policy.	Open homosexuality is informally severely sanctioned. Commanders and psychiatrists decide how to deal with homosexual behavior. No gay/lesbian support groups. Lesbians more acceptable. Homosexuals rarely "come out."

Country	Accessions		In Service	
	Policy	Practice	Policy	Practice
Italy (Conscription)	Homosexuality decriminalized in 1985. No specific policy regarding drafted homosexuals. Cross-dressing a criminal offense—seen as attempt to disguise one's identity. Homosexuality considered a behavioral anomaly, not pathology.	Homosexual draftees accepted, but not volunteers. Homosexual draftees can self-eliminate—antimilitary gay association helps them make the case—but few do this. Homosexual orientation is tolerated; homosexual behavior is not.	No specific policy.	Draftees who "come out" are usually released to avoid persecution by fellow soldiers. Junior officers and NCOs are usually discharged. Senior officers and NCOs are usually transferred.
Belgium (Volunteer)	No difference between civilian and military law in employment rights. Military is part of civil service. Sexual preference not asked.	Homosexuals are accepted—unless behavior associated is extreme. Transsexuals, transvestites, and those posing other problems are referred for medical exam.	No discharge for homosexuality. But sexual harassment or sex on ships results in transfer or, if associated with other disorders, discharge.	Open displays of homosexual behavior result in transfer from unit—commander's action. Homosexuals may be denied security clearance for some very sensitive jobs—i.e., they are considered more susceptible to blackmail.
Germany (Conscription)	Homosexual men are considered fit for military service.	Men are asked about homosexuality in the medical exam. Homosexuals are considered aptitude-deficient and mentally unfit for service. They are almost never admitted to military service or given alternative service.	Homosexuality not a reason for discharge. Homosexuals may not command or instruct troops. Open active-duty homosexual activity may be tried by disciplinary court.	Homosexuals considered unfit for military service and are usually discharged on those grounds. Officers usually discharged. Those with more than three years of service may receive a salary cut and/or ban on promotion.

Country	Accessions		In Service	
	Policy	Practice	Policy	Practice
Netherlands (Volunteer)	Strong gay movement led to legal changes in 1971. 1983 constitution change: No discrimination allowed (gays implied). 1974: homosexuals no longer rejected. 1978: gay military group founded.	Can self-eliminate by declaring homosexual status—must prove self unfit because of homosexuality; homosexual status by itself not sufficient.	"Emancipation" and integration are official policy. No discrimination in command or posting constraints.	Emancipation and integration are somewhat slowed by real-world constraints in acceptance of homosexuals. Published leaflets say homosexuals are welcome in the military.
Sweden (Conscription)	Since 1976 not a medical problem. Since 1984 can serve if coping is not a problem. Not asked about homosexuality during psychological interview—only about problems.	Few say they are homosexual at entry. To avoid service, must state that homosexuality is a problem to serving fully.	No discrimination in assignments or promotions.	Homosexuals serve openly. May be some discrimination in career opportunities/ promotions. Openly homosexual individuals might be harassed by peers.
Denmark (Conscription)	1954–78: registered service with home guard only, but could be disqualified by declaring homosexuality. After 1979: no questions—serve on equal terms with others.	Can self-eliminate by declaring homosexual status would be a service problem. No alternative service required.	No discharge for homosexuality. No HIV testing except for pilots when trained in the U.S. No posting or promotion constraints.	Very little "coming out."

Policies and Practices for Homosexuals' Military Service by Country (continued)

Country	Accessions		In Service	
	Policy	Practice	Policy	Practice
Norway (Conscription)	No discrimination. Not questioned about sexual orientation/behavior. Homosexual marriages/couples recognized.	Declaring homosexuality not sufficient to avoid service—must show behavior is/would be disruptive.	No discrimination in posting or promotions.	Very little "coming out."
Israel (Conscription)	1988: sodomy decriminalized. 1992: amended labor laws prohibit discrimination against homosexuals.	No attempt to identify homosexuals at entry.	No discrimination permitted. No policies to promote acceptance of homosexuals.	No apparent problems with new policy.

Sources: Segal, Gade, and Johnson (1994); Harries-Jenkins (1993 and in press); Kahan et al. (1993); and GAO (1993).

The second pattern shows a distinct relationship between the country's policy on sexual orientation and other areas of integration such as race, ethnicity, or gender. In most nations, Canada and the Netherlands in particular, sexual orientation integration has been seen explicitly as a civil rights or human rights issue.

Most nations do not categorically exclude homosexuals. Some of those that in the past have excluded homosexuals have changed their policies in recent years. We know of no nation that in the past has admitted homosexuals and has recently moved to exclude them. Thus, the number of nations that exclude homosexuals from military service has been declining.

Anglo-American Nations

Canada. We begin by examining our nearest neighbor, Canada. Canada may be the most relevant case for the United States, because the issue of admitting gay men and lesbians into the military services was resisted by the military and ultimately decided by the court system, and that appears to be the direction in which the United States is headed. Moreover, the United States and Canada share similar cultural values, a border, and, at least in part, a language. At the same time, we recognize that the military plays very different roles in Canadian and American societies.

The enactment of equal rights legislation under the Charter of Rights and Freedoms in 1978 and particularly in 1985 set the stage for lifting the ban on homosexuals in the military. In a 1985 review of the equal rights provision of the charter, the Canadian Justice Department cited the Canadian Forces as potentially being in violation of the charter in five areas: mandatory retirement ages, physical and medical employment standards, the recognition of common law relationships, the employment of women, and discrimination based on sexual orientation. This began six years of study and debate concerning the service of homosexuals (Pinch, 1994).

During this debate, issues of military cohesion and morale, confidence in the Canadian Forces, performance of the forces, discipline, privacy, equity, recruiting, retention, and medical concerns were studied. Expert testimony concerning all of these topics was given in hearings to assess the military impacts of allowing homosexuals to serve. Interestingly, much of this testimony was given by experts from the United States.

Surveys of the Canadian Forces about the impact of allowing homosexuals to serve were conducted by Zuliani (1986) and the Urban Dimensions Group, Inc. (1991). The results of those surveys showed that military personnel, especially men, were very much against lifting the ban on homosexuals, and predicted dire results if the ban were lifted. Zuliani found that personnel were negative about all aspects of service with respect to homosexuals. Most said they would refuse to share living accommodations with known homosexuals and many, particularly those in combat units, would refuse to work with openly homosexual men. Many heterosexual men said they would refuse to be supervised by homosexuals. Women in general were more accepting and tolerant of homosexuals than were the men in the study—a finding consistent with surveys in the United States and other countries.

In 1990, the Canadian Forces contracted with Urban Dimensions Group, Inc. to replicate Zuliani's survey. The new survey was to correct some of the procedural problems identified in the Zuliani study and to look for changes in attitudes over the four-year period. Unfortunately, a major sampling error that overrepresented service members from the Atlantic provinces crippled the new study. The sampling error, coupled with a very low 40 percent response rate, rendered it ineffective for generalizing to the Canadian Forces population (Pinch, 1994). It is worth a brief but guarded look, however, because it did show some consistency with the Zuliani survey. Once again, men were more hostile toward the idea of homosexuals serving than were women. A majority of respondents believed that homosexuals would be harassed in the military services and many said that the Canadian Forces would do nothing to protect heterosexuals or homosexuals from sexually harassing each other (UDG, 1991).

None of the evidence presented a clear, strong case for banning homosexuals, and at least one investigator, Lois Shawver, offered some evidence, although weak, that lifting the ban would not result in the dire events predicted by proponents of the ban (Pinch, 1994). The Canadian Federal Court concluded that there was no compelling evidence for the ban and determined that the Canadian Forces were in violation of the charter by restricting the military service of homosexuals. The Canadian Forces immediately complied with the court's decision and directed that the ban be lifted in 1992.

The implementation of a no-restriction policy began when restrictions on homosexuals were officially lifted by directives from the assistant minister for defence and from the chief of the defence staff. Explanations for the change were given, along with a reminder to the leaders that they were responsible for implementing the change. The defence leaders also appealed to the sense of fairness of service members to enlist their support for the policy. Finally, although not explicitly stated, those who could not support the policy were given the option of leaving the Canadian Forces.

The policy change was implemented as an equal rights, nondiscrimination issue. The approach was deliberately low-key and no special education or training focused on sexual orientation policy was proposed or conducted. Strict guidelines were issued for governing all heterosexual and homosexual interactions. Sexual misconduct and harassment, from whatever source, were not to be tolerated. All references to specifically homosexual conduct were removed from regulations and guidelines. Thus, all servicemembers were to be treated the same regardless of sexual orientation. No questions were asked of applicants or service members about their sexual orientation, and should orientation become known, it could not be used to restrict promotions, assignments, and so forth. Canadian Forces members cannot refuse to work or berth with other service members based on their sexual orientation.

Some issues were left unresolved. First, there was no plan for dealing with members who had been released from service for their sexual orientation prior to lifting the ban. It appears that these have been negotiated on a case-by-case basis. Second, the issues of homosexual marriages and partnership benefits have not been addressed, and Department of National Defence officials seem to have no plans to do so, apparently preferring to wait for the civilian government to resolve the issues (GAO, 1993, p. 30).

So far, the policy changes and their practical implementation seem to have progressed smoothly. None of the dire predictions about performance, recruiting, retention, and violent reactions has proved correct. Part of the reason for the successful implementation was the immediate and full support of the policy change by the leaders in the Department of Defence and in the Canadian Forces. They made it a leadership responsibility at every level of the Canadian Forces and showed they meant to implement the policy fully. Moreover, special attention was directed away from homosexuals rather than toward them as a special case. They were to be treated like everyone else, not like a protected class. Finally, a poor job market and economy and a need to reduce Canadian Forces personnel may have had a positive impact as well.

At any rate, the Canadian case appears to be a successful removal of a ban on homosexual service. The ultimate impact on the Canadian Forces is unknown.

The United Kingdom. The United Kingdom is a particularly interesting case because it highlights the direction of social change, reflects a frequently found divergence between official policy and actual practice, and exemplifies a commonly found pattern: that of limited tolerance. It is also the country most like the United States in policy and practice. In terms of official policy, when most homosexual acts were decriminalized in the 1967 Sexual Offenses Act, the British military was exempted from decriminalization. In May 1991, a parliamentary Select Committee on the Armed Forces recommended decriminalization in the armed forces as well, and in June 1992, the government accepted this recommendation. Although decriminalized, homosexuality is still regarded as incompatible with military service and is grounds for denial of enlistment or instant dismissal (Harries-Jenkins and Dandeker, in press; Harries-Jenkins and Dandeker, 1994).

Thus, the official policy is one of exclusion. As indicated in the table, however, military personnel are not asked about their sexual orientation. The practice is not to act unless they call attention to themselves. Indeed, if their orientation becomes known, but they are not openly engaged in homosexual behavior, they might be counseled and warned against misconduct, rather than discharged. If they are discharged, in the great majority of cases, the action is administrative rather than disciplinary.

Latin Europe

A common pattern in Latin Europe might be labeled laissez-faire. A revision of the military service laws in Portugal in 1989 set aside regulations prohibiting homosexuals. Spain decriminalized homosexuality in the military in 1984 (Burrelli, 1993). France takes a similar position.

France. Homosexuality is not a major issue for the armed forces, according to French officials and researchers (Kahan et al., 1993). Unless there are behavioral problems, the military regards sexuality as a private matter, and in fact homosexuality per se is not grounds for avoiding conscription, although behavioral problems associated with homosexuality may preclude service (Harries-Jenkins, in press-b; Robert and Fabre, in press). In practice, sexual orientation can and does make a difference. For example, if sexual orientation becomes known during the accession physical examination, the physician is likely to invite the examinee to request an exemption from service. Once in the military, conscripts may be given early medical discharges if they are found to be homosexual. This is rare, however. French officials prefer to ignore orientation unless behavior becomes a problem. Careerists' sexual orientation is ignored, again, unless it becomes a problem. Basically, the French do not tolerate sexual displays, whether heterosexual or homosexual, in military settings. This is a common characteristic of Latin European countries.

Italy. Officially, the Italian government does not discriminate or permit discrimination against homosexuals. In practice, however, Italian society is fairly unaccepting of homosexual behavior, considering it to be socially less acceptable than sexual relations between minors (Nuciari, in press). Homosexual behavior can be illegal if it results in enticement or prostitution, or if one disguises oneself in public—all of which are illegal regardless of one's sexual orientation. As in the rest of Latin Europe, homosexuality, like other sexual matters, is largely ignored if it is kept a private matter and does not result in scandal.

In Italy, until 1985, homosexuality was a basis for exemption from conscription. It is no longer a categorical reason for exclusion, and, as in most European societies, conscripts are not routinely asked their sexual orientation. But a conscript or enlistee may be discovered to be homosexual in the course of routine psychological examinations. In this case, the government makes a distinction between homosexual orientation and homosexual behavior. Homosexual orientation does not exclude one from military service, but homosexual behavior is considered to be a sexual deviation and will almost certainly result in an exemption from military service.

Individual conscripts may declare themselves as homosexual to try to avoid military service. To achieve this, they must convince the Military Medical Psychological Unit that there are or would be behavioral problems associated with their sexual orientation. Among career military personnel, as well as conscripts who enter the service, homosexuals are unlikely to reveal their sexual orientation, and if they are discovered, they are likely to be encouraged to resign, be discharged, or be transferred (Nuciari, in press).

Northern Europe

Belgium. Belgium is in the north, but is close to Latin Europe and has a significant Francophone population; like its Latin Europe neighbors, Belgium holds a laissez-faire position. There are no laws, rules, or regulations discriminating against homosexuals in the military, as long as they separate their personal and professional lives. Indeed, homosexuals have equal employment rights in both the civilian and military arenas (Manigart, in press). In this sense, Belgium is more explicitly tolerant than other laissez-faire countries and is closer to the dominant Northern European pattern. In the past, homosexuals were not allowed to serve in the para-commando regiment, but this appears to have been a function of the commander's preference rather than service policy. Soldiers whose sexual behavior is abusive of peers, either in the form of harassment or disruption of the unit, are subject to reassignment or medical discharge.

Germany. Germany is an especially interesting case because like Great Britain, it manifests a major difference between policy and practice, but in the opposite direction, and because it lies at one end of the range of patterns. It is also an extremely complex case, because its armed forces are composed of both conscripts and volunteers, who, as in the Italian case, serve under different policies, and because there are civilian service alternatives to conscription for conscientious objectors. In practice, as reflected in the table, Germany is the most exclusionary country we have studied.

Like most European countries, Germany does not officially regard sexual orientation as a relevant criterion for conscripted military service. But homosexual volunteers cannot be inducted, and career personnel who are discovered to be homosexual are limited in the nature of their service (GAO, 1993; Burrelli, 1993). Officially this is not because of homosexuality per se but because of its presumed impact on their military aptitude and leadership ability (Fleckenstein, in press). In practice very few homosexuals seem to serve. Unlike most nations, Germany asks conscripts and volunteers about their sexual orientation during the accession physical examination, as noted in the table, although this is not an official policy.

Many homosexual young men appear to choose alternative civilian service. Those who are conscripted, if they reveal their sexual orientation during in-processing, are likely to be rejected as "mentally unfit for service," thereby avoiding both military service and alternative civilian service. Official policy states that regular servicemen and volunteers are not rendered unfit by homosexuality, nor can they be discharged for a homosexual orientation. If they are discovered to be homosexual and have served for more than four years, they are not discharged before their term of service is completed. In practice, however, as noted in the table, if their orientation becomes known, homosexuals will not be allowed to assume supervisory positions or to serve as instructors. They may be restricted from high-security assignments as well. Junior officers within three years after their commissioning may be discharged on grounds that they are unfit for a career as an officer.

As in similar surveys of the military forces of other nations, German conscripts, in a 1992 survey, expressed low esteem for homosexuals. These conscripts found homosexuality less acceptable than unmarried cohabitation, prostitution, and abortion (Fleckenstein, in press).

Homosexuality has been decriminalized in German society, and homosexual behavior by military personnel off duty is not prosecuted. But the courts have affirmed the right of the military to prosecute soldiers for homosexual acts while on duty. Molesting a subordinate is grounds for discharge. Less serious offenses may be punished by demotion, banning promotions, and salary cuts.

The Netherlands. The Netherlands has probably the most tolerant position regarding homosexuals in the military. As Dutch sociologist Jan van der Meulen reported at a 1991 conference, "The acceptance and integration of women, ethnic minorities, and homosexuals in the armed forces was initiated as principle and policy before the end of the Cold War. . . . This does not mean that women, ethnic minorities, and homosexuals nowadays meet no discrimination any longer, nor that all three integration processes are concurrent" (Van der Meulen, 1991). More recently, Anderson-Boers and Van der Meulen have elaborated the Dutch position. They emphasized that sexual orientation has been ruled out as a selection criterion for both conscripts and career personnel, and that when problems occur, they are seen not as a consequence of homosexual orientation, but of homophobic reaction (Anderson-Boers, in press; Anderson-Boers, 1994).

Because the Netherlands is probably the most open and tolerant society with regard to homosexuality in the military, it has been in a position to undertake candid policy initiatives to make integration work, and to conduct research on problems of integration. Because of its belief that such problems are primarily a result of homophobia rather than of homosexuality per se, the military has instituted a training program to familiarize personnel with the range of homosexual lifestyles and to attempt to dispel incorrect stereotypes (Van den Boogaard, in press). In a major survey in 1991, a very small proportion of military personnel reported themselves to be homosexual or lesbian (about 1 percent of men, 3.5 percent of women). Most heterosexual personnel expressed tolerance for the rights of homosexuals, but 30 percent of male respondents indicated that they would respond in a hostile or aggressive manner if a colleague turned out to be homosexual, and about 60 percent of all respondents said that they preferred to maintain social distance from homosexuals. Known homosexuals are effectively excluded from social activities. Not surprisingly, most homosexuals in the service seem to prefer not to declare their sexual orientation. Research has documented the persistence of discrimination in the face of non-discriminatory policies (Ketting and Soesbeck, in press).

The defense minister has established a Commission for Advice and Coordination on Homosexuality in the Armed Forces, and homosexuals in the service have their own union, which is financially supported in part by the Defense Ministry.

Scandinavia
Although the Scandinavian countries are considered to be politically and sexually liberal, they have not always been as liberal as one might expect

in their treatment of homosexuals. For example, homosexual partnerships are legally recognized in all the Scandinavian countries. But Norway prohibits church sanctification of homosexual marriages (Kahan et al., 1993), as does Denmark, which also prohibits homosexual couples from adopting children (Hansen and Jorgensen, 1993). Although all three Scandinavian countries have slightly different approaches to homosexuals, all are very similar in their treatment of the issue of homosexuals in their military services.

Sweden. Sweden has a long history as an advocate for human rights and as a safe haven for individuals denied freedom in their own countries. Since 1987, discrimination against homosexuals has been illegal. Before 1979, homosexuals were automatically excluded from military service. In 1979, the military no longer classified homosexuality as an illness. In 1984, Sweden stopped asking recruits if they were homosexual, and in 1987 all restrictions on homosexuals were removed to comply with the newly enacted antidiscrimination law.

Sweden's policies and practices toward homosexuals in the military seem to be fairly consistent. Although homosexuals may avoid military service if they declare that their homosexuality would interfere with their ability to serve, this almost never happens. Although there are some hints of discrimination against careerists who are known to be homosexuals, Swedish society so effectively censors sexual matters that one's sexual orientation is almost never known by anyone else, nor do others seek to know it. In brief, sexual matters are a very private concern. Sexual matters are not discussed in public, and sexual displays, heterosexual or homosexual, are not permitted in or out of the military (GAO, 1993).

Denmark. Before 1955, the military did not ask about or monitor the sexual orientation of draftees or recruits. From 1955 to 1978, conscripts and volunteers were asked about their sexual orientation. Although homosexuals were registered for the draft, few ever entered the service. Normally they were told they could avoid military service if they could get a court order confirming their homosexuality. The national association of gays and lesbians protested this automatic rejection of homosexuals, and in 1979 the military returned to the policy of not asking about or using information about sexual orientation to exclude homosexuals from military service. In practice, however, gays can still escape military service if they declare that their homosexuality would interfere with their ability to serve.

The official policy of Denmark prohibits restrictions in promotions and assignments because of sexual orientation. This is followed in practice as well. Generally Danes do not want to know about anyone's sexual orientation or activities. They expect people to behave in a discreet manner, and as a result, sexual matters regarding military service are rarely a problem. Furthermore, service members almost never "come out," so the issue of homosexuals never becomes a problem for the military.

Norway. As in all the other Scandinavian countries, openly declaring one's sexual orientation and/or sexual practices is considered antisocial behavior. Military service by homosexuals was forbidden until 1979, when homosexuality was removed as a medical condition exempting one from service. Sodomy had already been decriminalized in 1972. As of 1993, Norway's official policy is to treat homosexuals the same as heterosexuals up to and including provision of housing and other spouse benefits to homosexual married service members. Homosexuals may still escape military service if they can substantiate claims that their sexual orientation or behavior is accompanied by psychological problems that would prevent them from serving. The exemption is not as freely given as it is in Denmark.

No problems are apparent with careerists who are homosexuals. They are treated equally and there is no violence against them. Officers are given regular training on matters of sexual orientation and behavior as part of their leadership training (Kahan et al., 1993). This training, together with the strong cultural censoring of all sexual behavior, provides a strong base for implementing Norway's policies.

Israel

Israel is perhaps the country of greatest interest to the United States, because it has fought so many wars effectively. It is here that the United States is most likely to look for answers to the questions about the effects of homosexuals on military cohesion and performance in a fighting army.

Homosexuals have always been allowed to serve in the military in Israel. But they were generally restricted to units where they could live at home, and were prohibited from occupying intelligence and other sensitive positions. In 1993, the Israeli Defense Force lifted all restrictions on homosexuals in the military service (Gal, 1994).

Practices seem to be in line with Israeli Defense Force policies (General Accounting Office, 1993; Gal, 1994). Homosexuals are not excluded from the draft, nor are they asked about their sexual orientation as part of the accession process. There is no evidence that careerists are restricted in assignments or promotion. When a soldier's homosexual orientation causes him to have a difficult time of adjusting to a unit, he may be moved to a new unit. This behavior, however, is treated no differently from any other type of adjustment problem.

Israeli society still has strong taboos against homosexual behavior, although people are very tolerant of it. Almost no one openly declares his or her sexual orientation. As it has in many other cultures and countries, this societal censoring of all public sexual behavior has probably prevented major problems from occurring in the military and other social organizations. That Israeli soldiers are used to mixed-sex housing has undoubtedly contributed to the lack of problems with the service of homosexuals.

Although the Israelis believe they have successfully integrated homosexuals into the military with no negative effects on cohesion and performance, they have not studied the impact, nor do they have any plans to do so. Moreover, the Israeli Defense Force does not conduct any training of its leadership or unit personnel about matters of sexual orientation, nor does it plan to do so (GAO, 1993).

Conclusions and Lessons Learned

Some nations, such as Germany, have policies of equality but practices of exclusion. Others, such as England, have policies of exclusion but practice limited tolerance of homosexuals in the military. Still others, such as France, Italy, and Belgium, have laissez-faire practices. But a few nations treat homosexuals as a privileged minority, at least in the military accession process. In the Scandinavian countries, for example, homosexuality was initially ignored by the armed forces in the post–World War II years. Until the late 1970s, draftees were asked about their sexual orientation, and homosexuals were registered and, in some cases, discharged. Yet draftees are no longer asked about sexual orientation and homosexuals are no longer registered. Homosexual draftees can avoid military service, with varying degrees of difficulty among countries, by claiming that their sexual orientation is psychologically incompatible with military service. Thus, whether the homosexual draftee serves is self-determined (Sorenson, in press).

Other Western nations offer our most useful current information about homosexuals and military service and our most promising societal laboratories for the future. We have found that the trend among Western cultures seems to be moving toward removing most discrimination, including restrictions on military service, against people because of their sexual orientation. Tolerance of homosexuality or national movements in that direction, rather than active integration, appears to be the cultural norm. The notable exception is the Netherlands, which has taken an activist role toward integrating homosexuals into society in general and the military in particular.

In most Western countries, including the Netherlands, homosexuals who serve in the military do so without revealing their sexual preferences. Heterosexuality is still the norm for the vast majority of those in Western nations and homosexuality in the military, as well as in other walks of life, is likely to remain shunned. As a result, homosexual "bashing" may be a particular problem for military services, especially in countries where bans against homosexuals in military service have only recently been removed. To derive lessons from these changes, it is imperative that social scientists continue to study countries that have recently removed restrictions on homosexuals in the military.

Although public opinion in the United States is becoming more tolerant of homosexuals, the vast majority of men in the military services are strongly opposed to homosexuals serving in the military (Miller, 1994; Segal, Gade, and Johnson, 1994). Women and African Americans in uniform are less strongly opposed, although they are more strongly opposed than their civilian counterparts (Segal, Gade, and Johnson, 1994).

We know from the surveys of Canadian, Dutch, and U.S. military personnel that there is the potential for violence and disruption if homosexuals serve openly (Zuliani, 1986; Urban Dimensions Group, Inc., 1992; Anderson-Boers and van der Meulen, 1994; Miller, 1994; Segal, Gade, and Johnson, 1994). But no violence has been reported in countries that have removed bans, even though surveys in those countries found that service personnel said that they would react violently to homosexuals in their units. Similarly, problems with cohesion, morale, recruiting, and retention have failed to materialize as predicted by soldier attitudes.

The failure of soldiers' negative attitudes toward homosexuals to predict actual behaviors may be an artifact of the measurement of those attitudes. These attitudes have been measured on a very general level, whereas the behaviors they might predict are much more specific with respect to action, target, context, and time. The ability of attitudes to predict behavior is thought to be maximized when all are measured at the same level (Eagly and Chaiken, 1993). Moreover, attitudes measured in surveys conducted so far have not accounted for the impact of social norms on regulating such behaviors. In current surveys, for example, references to violent behavior are vague as to specific action and context and do not assess what the effects of social norms in the military unit would be.

The failure of surveys of military personnel to measure attitudes appropriately may lie in their atheoretical construction. The surveys conducted so far have been of a purely pragmatic nature, with no attention to the theory of attitude-behavior linkage. To assess more realistically the likelihood of negative behavioral consequences of homosexuals serving in the military, we need to construct surveys that measure attitudes and intentions at the same level of generality and include some measure of social norms with respect to the behavior in question. Fishbein and Ajzen's theory of reasoned action (1975) has proved effective in constructing surveys that attempt to predict enlistment and retention behavior (Adelman, Pliske, and Lehner, 1987; Nieva, 1995). These seem to offer a viable approach here as well.

What, then, are some of the major inferences we can draw for the United States from our review of the experiences of other nations? First, it seems clear that, as in Canada and other Western countries, the issue of gays in the military will ultimately be determined by the nation's highest court. The "Don't Ask, Don't Tell" policy is difficult for commanders to implement. Cases challenging the policy are already winding their way rapidly through the lower court system, and conflicts with the new law continue to occur.

Second, if the military services are eventually ordered to cease excluding homosexuals who engage in homosexual behavior, they will do so quite effectively and without major incidents, provided that the leadership institutes policies and implements practices that clearly communicate support for the change. Moreover, the dire predictions of lost unit cohesion, violent disruptions, mass exodus from the services, and dismal recruiting are not very likely to come true. None of these things has happened as a result of the change to the "Don't Ask, Don't Tell" policy, and none of them has happened in the countries that totally lifted their bans on homosexuals. This probably has depended in part on the behavior of gay and lesbian organizations, however. If homosexuals or homosexual organizations engage in behaviors that draw negative attention to homosexuals within the military establishment, violence and major disruptions might erupt. On the other hand, if gay and lesbian organizations maintain a low profile, as they have done even in the most liberal societies we studied, then the inclusion of homosexual soldiers will likely occur without incident.

Third, and somewhat related to the above issue, those in the military services are likely to tolerate homosexuals, but we think they will be unlikely to accept them fully. None of the countries we studied, including the most liberal societies, fully accepted homosexuals and homosexuality. Given the attitudes of the general public regarding homosexuality, the United States is not likely to be an exception to this trend.

Fourth, even if the Supreme Court forces the military to include homosexuals, it seems unlikely that this will further liberalize society. Quite the contrary; it was the general rule in the countries we studied that society was liberalized first with respect to homosexuals and the military followed, not the reverse.

Finally, exclusion of one's sexuality, as well as race and gender, from the workplace appears to be emerging as a norm in Western societies that will regulate homosexual as well as heterosexual behavior in that context. Strong social censorship of public sexual behavior in the countries we studied appears to have had a dampening effect on many of the problems that might arise with the lifting of restrictions on the military service of homosexuals. The process is far from complete, however, and the path to resolution may be a difficult one in nations such as the United States, which have traditionally held strong exclusionary attitudes toward homosexuals in military service.

References

Adelman, L., Pliske, R. M., & Lehner, P. E. (1987). An empirical investigation as to the need for multicomponent decision models. *IEEE Transactions on Systems, Man and Cybernetics, 17,* 813–919.

Anderson-Boers, M., & Van der Meulen, J. (1994). Homosexuality and the armed forces in the Netherlands. In W. J. Scott & S. C. Carson (Eds.), *Gays and lesbians in the military: Issues, concerns, and contrasts* (pp. 205–216). New York: Aldine de Gruyter.

———. (in press). Homosexuality and armed forces in the Netherlands. In G. Harries-Jenkins (Ed.), *Comparative international military personnel policies.* Alexandria, VA: U.S. Army Research Institute for the Behavioral and Social Sciences.

Burrelli, D. F. (1993). *Homosexuals and U.S. military personnel policy* (93-54F). Washington, DC: Congressional Research Service.

Eagly, A. H., & Chaiken, S. (1993). *The psychology of attitudes.* Fort Worth, TX: Harcourt Brace Jovanovich College Publishers.

Fishbein, M., & Ajzen, I. (1975). *Belief, attitude, intention, and behavior: An introduction to theory and research.* Reading, MA: Addison-Wesley.

Fleckenstein, B. (in press). Homosexuality and military service in Germany. In G. Harries-Jenkins (Ed.), *Comparative international military personnel policies.* Alexandria, VA: U.S. Army Research Institute for the Behavioral and Social Sciences.

Gal, R. (1994). Gays in the military: Policy and practice in the Israeli defence forces. In W. J. Scott & S. C. Carson (Eds.), *Gays and lesbians in the military: Issues, concerns, and contrasts* (pp. 181–189). New York: Aldine de Gruyter.

General Accounting Office. (1993). *Homosexuals in the military: Policies and practices of foreign countries.* (GAO/NSIAD-93-215). Washington, DC: Author.

Hansen, B., & Jorgensen, H. (1993). The Danish partnership law: Political decision making in Denmark and the national organization for gays and lesbians. In A. Hendriks, R. Tielman, & E. van der Veen (Eds.), *The third pink book: A global view of lesbian and gay liberation and oppression* (pp. 86–99). New York: Prometheus Books.

Harries-Jenkins, G. (1993). *Comparative international military personnel policies* (Research Note 93-17 (NTIS No. ADB 154547)). Alexandria, VA: U.S. Army Research Institute for the Behavioral and Social Sciences.

———. (in press-a). Final research report: Comparative international military personnel policies. In G. Harries-Jenkins (Ed.), *Comparative international military personnel policies.* Alexandria, VA: U.S. Army Research Institute for the Behavioral and Social Sciences.

———. (in press-b). The military and homosexuality: The French case: A summary. In G. Harries-Jenkins (Ed.), *Comparative international military personnel policies.* Alexandria, VA: U.S. Army Research Institute for the Behavioral and Social Sciences.

Harries-Jenkins, G., & Dandeker, C. (1994). Sexual orientation and the military: The British case. In W. J. Scott & S. C. Carson (Eds.), *Gays and lesbians in the military: Issues, concerns, and contrasts* (pp. 191–204). New York: Aldine de Gruyter.

———. (in press). Homosexuals in the armed forces of the United Kingdom. In G. Harries-Jenkins (Ed.), *Comparative international military personnel policies.* Alexandria, VA: U.S. Army Research Institute for the Behavioral and Social Sciences.

Herek, G. M. (1993). Sexual orientation and military service. *American Psychologist, 48,* 538–549.

Kahan, J. P., Fulcher, C. N., Hanser, L. M., Harris, S. A., Rostker, B. D., & Winkler, J. D. (1993). Analogous experience of foreign military services. In National Defense Research Institute, *Sexual orientation and U.S. military per-*

sonnel policy: Options and assessment (pp. 65–105). Santa Monica, CA: RAND.

Ketting, E., & Soesbeck, K. (in press). Homosexuality and the armed forces in the Netherlands. In G. Harries-Jenkins (Ed.), *Comparative international military personnel policies.* Alexandria, VA: U.S. Army Research Institute for the Behavioral and Social Sciences.

Manigart, P. (in press). Homosexuals in the Belgian Army. In G. Harries-Jenkins (Ed.), *Comparative international military personnel policies.* Alexandria, VA: U.S. Army Research Institute for the Behavioral and Social Sciences.

Miller, L. (1994). Social science research on homosexuals in the military. In W. J. Scott & S. C. Carson (Eds.), *Gays and lesbians in the military: Issues, concerns, and contrasts* (pp. 69–85). New York: Aldine de Gruyter.

Nieva, V. F., Wilson, M. J., Norris, D. G., Greenless, J. B., Laurence, J., & McCloy, R. (1995). *Enlistment intentions and behaviors: A youth and parental model.* Unpublished manuscript.

Nuciari, M. (in press). Homosexuality and armed forces in Italy. In G. Harries-Jenkins (Ed.), *Comparative international military personnel policies.* Alexandria, VA: U.S. Army Research Institute for the Behavioral and Social Sciences.

Pinch, F. C. (1994). *Perspectives on organizational change in the Canadian Forces* (Research Report 1657 (NTIS No. ADA277746)). Alexandria, VA: U.S. Army Research Institute for the Behavioral and Social Sciences.

Robert, G., & Fabre, L. M. (in press). Armée et homosexualité: "The French agreement." In G. Harries-Jenkins (Ed.), *Comparative international military personnel policies.* Alexandria, VA: U.S. Army Research Institute for the Behavioral and Social Sciences.

Segal, D. R., Gade, P. A., & Johnson, E. M. (1994). Social science research on homosexuals in the military. In W. J. Scott & S. C. Carson (Eds.), *Gays and lesbians in the military: Issues, concerns, and contrasts* (pp. 33–51). New York: Aldine de Gruyter.

Sorenson, H. (in press). Homosexuals in the armed forces in Scandinavia. In G. Harries-Jenkins (Ed.), *Comparative international military personnel policies.* Alexandria, VA: U.S. Army Research Institute for the Behavioral and Social Sciences.

Urban Dimensions Group, Inc. (1991). *Canadian Forces (CF) internal survey on homosexual issues.* Toronto: Department of National Defence.

Van den Boogaard, H. (in press). Training courses on homosexuality for armed forces personnel. In G. Harries-Jenkins (Ed.), *Comparative international military personnel policies.* Alexandria, VA: U.S. Army Research Institute for the Behavioral and Social Sciences.

Van der Meulen, J. (1991, October). *The Netherlands.* Paper presented at the Inter-University Seminar on Armed Forces and Society and The Olin Foundation Conference, Baltimore.

Zuliani, R. A. (1986). *Canadian Forces survey on homosexual issues.* Ottawa: Charter Task Force, Department of National Defence.

Lessons Learned from the Experience of Domestic Police and Fire Departments

Paul Koegel

It is every policymaker's dream to discover a crystal ball with which to view the consequences of a policy change before implementing it. Because that crystal ball remains elusive, those with the responsibility of evaluating the potential impact of a new policy are left with little choice but to rely on the less perfect divining tools at their disposal. One such tool involves exploring analogies—that is, examining the implementation of similar policies in parallel situations to obtain insights on how a change might unfold. Such exercises allow glimpses into the future, but they carry risks as well. In that it is virtually impossible to identify the perfect analogy, the picture that emerges through examination of the closest possible analog may be distorted. Careful consideration of the relevance of an analogy and where it may or may not be useful is thus imperative.

Chapters 4, 5, and 6 use different analogies to explore issues related to permitting homosexuals to serve openly in the United States military. The first two analogies—those focusing on women and minorities—are strengthened because they explore the aftermath of the inclusion of out-groups into the U.S. military itself, but are limited by the possibility that there are significant differences between the out-groups they examined

This chapter is drawn from chapter 4 in *Sexual orientation and U.S. military personnel policy: Options and assessment* (1993), Santa Monica, CA: RAND Corporation. It is based on research conducted by Janet Lever, Brent Boultinghouse, Scott A. Harris, Joanna Z. Heilbrunn, James P. Kahan, Paul Koegel, Robert MacCoun, Peter Teiermeyer, John D. Winkler, and Gail L. Zellman.

and homosexuals. The third analogy—an examination of the experiences of foreign militaries—is strengthened by its focus on homosexuals themselves, but is limited because of differences between the United States and foreign militaries and because the United States and its foreign counterparts each have distinctive cultures, particularly with regard to privacy and social values.

Clearly, a fourth analogy—one that examines homosexuals within the U.S. cultural context in organizations that are as analogous to the military as possible—is needed to complement the preceding three. This chapter takes advantage of the similarities between municipal public safety departments and military organizations to describe the experience of police and fire departments in six American cities that have implemented policies of nondiscrimination based on sexual orientation. This is done with two primary purposes in mind: (1) to better understand what happened in these departments when policies of nondiscrimination were implemented, and (2) to better understand how the implementation process itself occurred. Overall, this chapter explores whether and how the experience of these paramilitary organizations sheds light on issues related to lifting the ban against homosexuals in the armed forces.

Although strengthened by its focus on homosexuals in a U.S. cultural context, this analogy, of course, is not without its own limitations. Its weakness lies in the use of police and fire departments as a proxy for the military. Police and fire departments are *not* identical in nature to the military. The members of the police and fire departments interviewed as part of this study were themselves quick to point out fundamental differences between their organizations and the armed forces. Most significantly, their members are on duty for short stints—an eight-hour shift in the case of police, a period of one to three days in the case of firefighters. Afterward, they go home, where they have far greater latitude in how they must behave. The military, on the other hand, takes service members away from their homes for extended periods of time for both training and deployment, and considers the boundaries of their jobs to be twenty-four hours a day, seven days a week. During that time, it often demands that service members live in a variety of close quarters, from the open dormitories of basic training barracks to the cramped confines of a two-person pup tent. Moreover, it requires them to subject themselves to the military and its codes of behavior at all times.

Even so, there are a number of characteristics that police and fire departments share with the U.S. military that make them the closest possible *domestic* analog (see Goldsmith and Goldsmith, 1974; Martin, 1980; Leinen, 1993; Niederhoffer and Blumberg, 1970; NIMH, 1970). Among them are the following:

- each is hierarchically organized with a well-defined chain of command; the uniforms carry insignia denoting rank;
- in each, the occupations are defined as public service for the maintenance of public security;
- members work together as teams and wear uniforms clearly identifying them with the organization;
- a substantial proportion of job time is spent training for short, intense periods of hazardous activity. An inherent feature of the job is putting one's life at risk.

Fire departments are characterized by even greater similarities with the military than are police departments. Firefighters typically live together in a firehouse while on duty, sometimes for days at a time. Close living quarters and issues related to privacy, especially in older firehouses, are thus part of their experience, even if for shorter periods of time. The work of fighting fires is done in coordinated fashion against a common enemy (as opposed to police work, which involves working in pairs or alone, and in which the provision of human services is a primary task). As in the military, the business of a fire-fighting company is tactical; the command structure concerns itself with the strategic allocation of resources.

All of this suggests that although there are significant differences between domestic paramilitary organizations and the military that limit the explanatory power of this analogy, there are also marked similarities that leave it well suited to addressing many important questions. These similarities may not be sufficiently strong to allow predictions related to privacy issues, cohesion or performance. But the analogy is well suited to answering questions regarding how many members of such organizations publicly acknowledge their homosexuality when a policy change occurs; the factors that influence this; the behavior of lesbians and gay men under a policy that allows them to acknowledge their homosexuality; the concerns that heterosexuals express after, rather than before, such a change has occurred; the role of leadership and chain of command; the natural evolution of policy implementation over time; and many other issues. It is on these issues that this chaper focuses.

In exploring the experiences of homosexuals in domestic police and fire departments, this chapter delves into somewhat uncharted territory.

Table 7.1. Cities Visited

City	U.S. pop. rank	Year Policy Changed
Chicago	3	1988
Houston	4	1990–1991
Los Angeles	2	1979
New York	1	1979
San Diego	6	1990
Seattle	21	1980

The attempt to integrate women and minorities into these public safety organizations has been well documented in a number of works (e.g., Alex, 1969; Craft, 1971; Heidensohn, 1992; Martin, 1980; Parachini, 1994). The literature on gay and lesbian police and firefighters, however, has largely been restricted to human interest stories that periodically appear in the popular press. Only recently have gay and lesbian police officers begun to receive more scholarly attention, although these examinations have been restricted to settings outside the United States (Burke, 1993; 1994) and to the New York City Police Department (Leinen, 1993), which has the largest concentration of known homosexual officers. Police officers outside these settings have generated far less attention. Virtually nothing of a scholarly nature has been written about the experience of gay and lesbian firefighters.

Methods

This examination of police and fire departments took place in six of the twenty-five largest cities in the nation. Large cities with sizable departments were selected to increase the odds of finding homosexuals who were serving in these organizations. Each of the nation's major regions was represented and only those cities that had a policy of nondiscrimination against homosexuals were targeted. All but two of the twelve selected departments agreed to participate in the study. Table 7.1 presents the six cities in which data collection occurred, along with their population rank and the year of introduction of a policy change. Table 7.2 presents some demographic information about these cities and their police and fire departments.

The data reported here were collected during a two-day visit to each city. During these visits, several data collection methods were utilized. The principal source of information consisted of interviews with key persons in the implementation process. Using open-ended interview techniques but guided by a detailed set of topic questions that was first piloted in the police and fire departments of Santa Monica, California, interviews were conducted with high-ranking leaders, personnel and equal employment officers, trainers, unit commanders, recruiters, and counselors. Although none of these interviews was recorded, for fear of inhibiting the free exchange of ideas on sensitive topics, extensive notes were taken— as close to verbatim as possible—at each. Interviews were conducted also with heterosexual rank and file members of the force and gay and lesbian members of the force, both alone and in a focus group format (Krueger, 1989; Stewart and Shamdasani, 1991), in groups ranging from three to twenty. Rank and file officers were recruited by department leaders for these one-on-one interviews and focus group discussions, usually based on who was available at the time set aside for the interviews, and were interviewed without leaders being present. Interviews with gay and lesbian force members usually took place during off-duty hours in off-site, confidential locations. In addition to involving individuals who had publicly proclaimed their homosexuality in the workplace, these meetings often included police officers and fire fighters who had not disclosed their orientation to their departments. Again, these were not recorded, and the notes excluded any identification of participants.[1] In addition to interview data, available documentation on the size and composition of the police and fire departments was gathered, along with policies and regulations

Table 7.2. Selected Demographic Information About Cities Visited

	Chicago	Houston	Los Angeles	New York	San Diego	Seattle
Population (in thousands)	2,784	1,631	3,485	7,323	1,111	516
% white	45	53	53	52	67	75
% black	38	28	14	29	9	10
% Hispanic	20	28	40	24	21	4
Uniformed police	12,200	4,100	7,700	28,000	1,800	1,300
% women	17	N/A	14	14	13	10
% minority	35	N/A	41	26	40	N/A
Uniformed fire	4,700	2,900	3,200	11,300	850	975
% women	4	0.6	N/A	0.3	8	7
% minority	28	27	N/A	6	28	24

Source: Census figures from *World Almanac* (1992); personal communications.

Note: The sum of Population percentages can exceed 100 because the census separately categorizes race and Hispanic origin.

Table 7.3. Sources of Information, By City

	Chicago	Houston	Los Angeles	New York	San Diego	Seattle
Police interviews						
Leaders	x		x	x	x	x
Personnel, EEO	x		x	x	x	x
Trainers	x		x	x	x	x
Commanders	x			x		
Recruiters				x		
Counselors	x					
Gays and lesbians	x	x	x	x	x	x
Rank and file	x			x	x	
Fire interviews						
Leaders	x			x	x	x
Personnel, EEO	x			x	x	x
Trainers		x			x	
Commanders	x	x			x	
Recruiters		x				
Counselors						
Gays and lesbians	x			x	x	x
Rank and file		x	x	x	x	x
Documentation						
Nondiscrimination policy	x		x	x	x	x
Police regulations, procedures	x		x		x	x
Police training programs	x		x			
Fire regulations, procedures	x	x		x		
Fire training programs						x
Newspaper articles	x	x	x	x	x	x

regarding nondiscrimination, enforcement guidelines, curricula for training programs, and equal employment procedures. Newspaper articles concerning events related to the implementation of nondiscrimination policies were also reviewed.

Not all investigative methods were employed at all sites. In each case, as much information was gathered as time and the goodwill of the organization allowed. Table 7.3 summarizes the types of information obtained from each department.

Consequences of a Nondiscrimination Policy

This section describes the consequences that followed the introduction of policies making it possible for acknowledged homosexuals to serve in police and fire departments, focusing first on the behavior and responses of homosexuals, including the number and characteristics of people who come out, the factors that influence this process, and the nature of their experiences. The second focus was on the attitudes and behavior of heterosexuals, including whether they accept lesbians and gay men and the nature of their concerns regarding working with acknowledged homosexual colleagues. These issues have been highlighted in public discussions of the impact of lifting the ban against homosexuals in the military and are among those that this particular analogy is perhaps best suited to inform.

The Experiences and Responses of Homosexuals

To what extent do lesbians and gay men acknowledge their homosexuality once a policy change occurs? Across all of the departments examined as part of this effort, exceedingly few lesbians and gay men announced their homosexuality despite the existence of policies that codify their right to serve (see Table 7.4). This was especially pronounced in fire departments; no currently active gay male firefighter had acknowledged his homosexuality and acknowledged lesbians were found in only two. Although there was general awareness that far more homosexuals were serving than were officially known in each of the departments, in no department did the percentage of openly gay and lesbian officers exceed 0.5 percent; the median value was 0.03 percent of the total force. Overall, gay and lesbian officers remained overwhelmingly

Table 7.4. Numbers and Percentages of Open Homosexuals in the Police and Fire Departments of Six Cities

Institution	City	Total Force Size	Number of Open Homosexuals	Estimated Prevalence (percentage)
Police	Chicago	12,209	7	0.06
	Houston	4,100	0	0.00
	Los Angeles	7,700	7	0.09
	New York	28,000	~100	0.36
	San Diego	1,300	4–5	0.25
	Seattle	1,300	2	0.15
Fire	Chicago	4,700	0	0.00
	Houston	2,900	0	0.00
	Los Angeles	3,200	0	0.00
	New York	11,300	0	0.00
	San Diego*	845	1	0.12
	Seattle*	975	5	0.51

*All openly homosexual firefighters in these cities were women.

reluctant to allow their homosexuality to become public knowledge, even when leaders in their departments were actively encouraging them to declare themselves.

That so few homosexuals had publicly disclosed their sexual orientation within their work setting was not a function of their being no more than a handful of homosexuals in each of these departments. Far more homosexuals than had publicly disclosed their sexual orientation were known to each other and to selected heterosexual members of their departments. In one department, for instance, only seven individuals had acknowledged their homosexuality to their department, but more than forty belonged to a confidential gay fraternal organization of department members. Moreover, in every city, gay and lesbian officers knew of other homosexual members of the force who had decided not to join such groups, either for fear of being identified as gay or for lack of interest. There is no way of estimating precisely how many homosexuals are serving in these departments, because people can keep their sexual orientation hidden. It is thus impossible to estimate what proportion of homosexuals declare their orientation. Still, it is clear that in these departments only a minority of homosexual members elect to disclose the fact of their homosexuality despite their legal ability to do so.

Perhaps one of the most salient factors that influences whether homosexual police officers or firefighters make their sexual orientation known to their departments is their perception of the *climate* in which they work. A marked degree of variation was apparent both between and within each of the departments in the messages communicated to gays and lesbians regarding the reception they would receive if they acknowledged their homosexuality. This variation could be observed along many dimensions—across and within the hierarchical levels of an organization; between police and fire departments (the latter being far less tolerant); across gender lines (with women generally being less likely than men to view homosexuality as offensive, troublesome, and threatening); and between gay men and lesbians (the climate with regard to lesbians being consistently more tolerant than with regard to gay men). Gay and lesbian officers made it clear that they carefully attend to the messages they receive on each of these levels. In general, the more hostile the environment, the less likely it was that people publicly acknowledged their homosexuality.

Variation in degree notwithstanding, data collected as part of this study indicate that most of these police and fire departments can be characterized as being overtly, and in some cases extremely, hostile toward homosexuals. Hostile attitudes, even in departments in which attempts at change were being actively pursued, left lesbians and gay men with serious fears about the consequences of revealing their homosexuality. They worried about their safety, their careers, and their social acceptance within the workplace. Given the prevailing attitudes in their departments, the majority believed that not going public was preferable to acknowledging their homosexuality. Thus, most kept their sexual orientation to themselves or shared the fact of their homosexuality only with highly trusted coworkers.

What are the actual experiences of those who do acknowledge their homosexuality? Paradoxically (given the high level of antihomosexual sentiment expressed in these departments), most people who publicly acknowledged their homosexuality reported that the consequences of doing so were far less dire than they or their unacknowledged counterparts feared. Each faced some degree of hostility, but this typically took the form of offensive remarks or epithets. Pranks were occasionally reported, but (with rare exceptions, as in the case cited below) gay and lesbian officers could rely on backup and faced almost no risk of overt violence from their colleagues. Most were socially accepted and even applauded for their courage; when they were not, social disruptions did not interfere with their doing an effective job. Many spoke of the frustration of having to prove themselves with each transfer to a new assignment, but most had confidence in their ability to do so, and believed that acknowledging their sexual orientation had enabled them to perform their duties more effectively. Many believed it improved their work environment, because people who had previously felt comfortable expressing antihomosexual sentiments curbed those remarks in their presence.

The experiences of these officers may seem to contradict the assertion made earlier that a climate of hostility toward homosexuals exists in these departments. Actually, it does not. Lesbian and gay men tend to come out in precincts in which hostility is less pronounced. They also tend to come out after they have proved themselves to be good officers (see also Leinen, 1993, p. 80), allowing them to be defined by those who retain antihomosexual feelings as the "exception to the rule." Finally, the antihomosexual sentiment evident in these departments often takes the form of negative remarks regarding homosexuality and homosexuals. These are not necessarily related to how these officers will *behave* in front of someone they know, though homosexual officers who have not disclosed their sexual orientation are not usually convinced of this.

Isolated examples of more serious and threatening hostility were occasionally reported. One officer, who had generally been viewed as a model policeman on the fast track before knowledge of his homosexuality became known, ultimately left his department and filed suit against it after a protracted series of incidents left him fearing for his life. Fellow officers engaged in hostile pranks, such as scratching threatening messages into his car; solicited a false accusation from a suspect that the gay officer had inappropriately strip-searched him; and ultimately failed to respond adequately to calls for backup. Another well-respected police officer recently left his department, citing his unwillingness to cope with daily affronts to his dignity any longer. Dire consequences, however, appeared to be the exception, rather than the rule. Interestingly, in the most serious instances of abuse against acknowledged gay or lesbian officers, the situation was usually one in which the officer's homosexuality had become public knowledge not by design but by accident—people had been "outed," in other words, or were merely suspected of being gay in departments in which an especially hostile climate toward homosexuals prevailed.[2] When homosexual officers were allowed to exercise their own judgment about the advisability of coming out, problems, if they emerged, were usually manageable.

Do acknowledged homosexual police officers and firefighters engage in personal behaviors that are disruptive to their organizations? Among those anticipating the inclusion of homosexuals in work settings such as the military or police and fire departments, it is an often-cited fear that lesbians and gay men will behave in ways that will challenge local institutional norms and customs by engaging in such practices as dancing together at departmental functions or sexually harassing heterosexual members of the force. Evidence to support these fears was very rare. Generally speaking, gay and lesbian officers are sensitive to the climate in which they work. Although there are occasional exceptions, the vast majority behave in ways that are designed neither to shock nor to offend. Not a single case of a gay male sexually harassing a heterosexual male was reported. Occasional reports were offered by commanding officers of lesbians harassing heterosexual women—staring at them in the locker room or making unwelcome sexual comments. These were said to be far more rare than incidents of heterosexual men harassing women. Public displays of affection were even more unusual. A few officers reported bringing same-sex partners to social functions, but only when it had been assumed that this would be accepted or would serve as a nudge, rather than a hard push, against the established social order.

Another way in which the behavior of gay and lesbian police officers and firefighters might inadvertently strain their organizations relates to how they react to the sometimes-daily instances of interpersonal harassment they face. A predisposition to file formal complaints regarding each incident of harassment could quickly overwhelm the systems in place to deal with these problems and exact further demands on scarce resources. In reality, formal complaints are rare. Homosexual officers appeared to have internalized norms evident within both police and fire departments regarding working out problems within the ranks and not informing on a peer. Most develop thick skins and either ignore or deflect the harassment they experience. Those who turn to the chain of command tend to do so informally, reaching out to a supervisor for assistance on the condition that he or she keep the complaint confidential. Usually, the goal is to end or contain the offensive behavior, not to punish the offending party. Formal complaints are invariably acts of desperation and are usually leveled only against those whose harassment is quite extreme.

What are the characteristics of lesbians and gay men who join police and fire departments? Many who contemplate the impact of opening military and paramilitary organizations to gays and lesbians worry that stereotypic homosexuals, particularly effeminate men, will compromise the image of their force. This was not the case in the departments included in this study. Homosexual individuals in these departments were virtually indistinguishable from their heterosexual peers. Almost unilaterally, gay men were reported as being, and seemed to us to be, sufficiently innocuous in their behavior and appearance to have been able to pass as heterosexual members of the force for long periods of time. Lesbians also tended to be indistinguishable from their heterosexual counterparts. Occasional stories were told by straight police officers of lesbians who came across as somewhat "butch," but this was said to work in their favor both on the beat and while socializing with the "boys" in the precinct houses. The characteristics of those gay and lesbian officers interviewed as part of this study, and the comments of heterosexual officers who were interviewed, suggested that those drawn to police work and fire fighting were unlikely to match stereotypes that were inconsistent with the job at hand.

In addition to resembling physically and behaviorally their heterosexual counterparts, gay and lesbian police officers and firefighters mirrored their heterosexual peers in terms of what attracted them to their work. No one interviewed as part of this study entered his or her departments to advance a homosexual agenda. When job-related passion was expressed, it tended to reflect a stronger identification with being a police officer or a firefighter than a member of the gay community.

As for performance, there was no question that gay members of these departments could do their jobs adequately. If anything, there was a sense among both leadership and patrol officers that homosexuals who have publicly acknowledged their sexual orientation tend to be overachievers. In part, this was attributed to the constant demand imposed on homosexual officers—both by themselves and by others—to prove themselves. In part, this was because it was thought among homosexual officers that only an untarnished record could allow known gays and lesbians to advance within the ranks.

The Responses and Concerns of Heterosexuals

To what extent do heterosexual police officers and firefighters accept homosexuals who acknowledge their sexual orientation? As indicated earlier, many heterosexual members of these forces harbor negative attitudes toward homosexuals that do not disappear once a policy of nondiscrimination is enacted. Overall, the prevailing climate across all the departments was hostile toward gay men and lesbians. Even so, among those who have actually worked with gays, there are signs of more accepting attitudes that, according to those in leadership, have been growing steadily over time. Many heterosexual officers in a variety of contexts stated that a person's sexual orientation was immaterial to them. Both straight and gay officers confirmed that gays were frequently, even if not consistently, part of off-duty social activities. Gay men and women pointed to the support they received from individual colleagues when they acknowledged their homosexuality and spoke of their surprise at both the strength and, in some cases, the source of that support. Some told stories of support from heterosexual coworkers who reassured them of their own comfort with the person's sexual orientation but warned them that others would have a hard time, only to hear those others say the same thing. Such heterosexuals, in other words, endorsed the notion of pervasive antihomosexual attitudes in their departments but saw themselves as an exception to that rule.

Even heterosexual officers who expressed less positive attitudes toward their gay and lesbian colleagues often adhered to a strong ethic of professionalism that allowed them to work smoothly with homosexuals. Among officers who had actually worked with homosexuals, the prevailing sentiment seemed to be that getting the job done was paramount. Even where expressions of antihomosexual sentiment were typical and an overall climate of hostility existed in the department at large, the majority of heterosexual officers interviewed insisted that whatever personal animosity they felt toward gays did not interfere with their interactions with them in the field, their mission or the overall goals of their department.

The apparent contradiction between descriptions of the antihomosexual climate of these departments provided to us, and the positive experiences reported by some of the acknowledged homosexual officers, suggests that the attitudes and behaviors of heterosexual members of these departments are complex and sometimes counterintuitive. Whereas strong negative and positive messages were evident to varying degrees across and within departments, much of what these officers expressed defies simplistic categorization. It was not unusual for officers to make seemingly contradictory statements or behave in ways that contradicted their stated attitudes, both negative and positive. Heterosexual officers themselves noted that many individuals who actually practice a "live and let live" credo were less likely to acknowledge this within the context of a group discussion. Similarly, officers repeatedly provided examples of how people actually behaved differently from their stated beliefs. There was also a disjuncture between the stated attitudes of heterosexuals and the actual experiences of many homosexuals who disclosed their orientation, and a clear difference in attitudes between those who had and had not actually worked with individuals who had publicly acknowledged their homosexuality. These points suggest that actual behavior cannot necessarily be predicted by stated attitudes, no matter how vitriolic they seem.

What concerns are voiced by heterosexual police and firefighters, particularly those who have had experience with gay colleagues? That police officers do not live together and that so few firefighters had disclosed their homosexuality make this analogy weak with regard to privacy issues, one of the major concerns raised in discussions of lifting the ban against homosexuals in the armed forces (see chapter 11). The analogy is not as weak, however, with regard to a second concern that emerged from interviews with heterosexual members of these departments: the risk of contracting the human immunodeficiency virus (HIV). In many cases, this concern was mitigated by training, but overall the fear of HIV remained high. Some police officers raised the question of whether they would provide emergency first aid to fellow officers known to be gay. Some firefighters expressed fears that exposure to the virus through shared dishes or use of bathrooms might put them at risk, implying that HIV might be more easily transmitted than common knowledge would have one believe. In one department, a lawsuit had been filed by an HIV-positive firefighter who agreed to take a detail outside a firehouse after knowledge of his HIV status became public, but subsequently claimed to have been coerced into doing so. The hysteria generated by this incident left department leaders believing that without the HIV issue, gay men could be integrated into firehouses without threatening operational effectiveness, but that given the high risk of contracting Acquired Immune Deficiency Syndrome (AIDS) among gay men, problems would be unavoidable.

Perhaps the most sharply expressed concern on the part of rank and file members of these departments, however, was the fear that gays and lesbians would achieve—indeed, in some instances had achieved—special class status. This issue spontaneously emerged in each of the focus groups conducted with heterosexual rank and file officers, most of whom were white and male. Outrage was consistently voiced at the possibility that gays and lesbians might be disproportionately hired, receive special promotional opportunities, be held to a lower standard, or be afforded special class protections (such as unique procedural pathways for lodging complaints). These individuals already believed that they were inhibited in their interactions with minorities and women because of the perception that such individuals could lodge formal complaints against them regarding behavior they themselves felt was harmless—that these groups had power over them because of the special protection they enjoyed under the law. They also perceived themselves as experiencing the sting of reverse discrimination with regard to women and minorities within their organizations and bitterly resented it. The last thing they wanted to see was another protected class.

To what extent are negative attitudes toward gays and lesbians subject to change? In police and fire departments alike, both the leadership and the rank and file members believed that change *was* occurring, albeit slowly, with regard to the attitudes of heterosexual officers and fire-fighters toward homosexuals. Many predicted that in twenty years far more homosexuals would acknowledge their sexual orientation and that many of today's seemingly intractable problems would be solved.

Several factors were cited as influencing attitudinal change. The inclusion of younger, better-educated cohorts of officers with more tolerant views of homosexuality was repeatedly mentioned, as was a general tendency toward social values becoming more liberal with time. "You constantly hear macho people saying, 'I'm not going to tolerate gays in the firehouse,'" one fire chief offered. "In the '60s, people claimed that they wouldn't sleep in a room with black guys, and look at things now. Things evolve and take care of themselves." Also mentioned was the process that elevates one's status as a police officer or fire fighter to a higher level of importance than one's status as homosexual, a transformation that usually occurred after a particularly competent or heroic handling of a dangerous situation. But *positive contact* was pointed to as by far the most potent determinant of attitudinal change. Given the opportunity to know gay and lesbian colleagues and thereby test the stereotypic images of homosexuals that many heterosexuals hold, heterosexual men and women could arrive at a different understanding of homosexuality. Gay officers concurred that contact could be the pivotal factor in turning around negative attitudes. "Most people don't know someone who is gay," one officer noted. "Once they get to know someone who is gay, the negative attitudes and behaviors start to break down. People are amazed to find out you have a full, well-formed life with a stable partner, and that you're not just out looking for anonymous sex. It's not being able to be honest that allows the stereotypes to continue."

There was far less consensus on the issue of whether formal sensitivity training facilitated attitudinal change among heterosexual officers. Gay and lesbian members of these departments tended to be strong advocates of training, believing that ignorance would give way to knowledge and understanding if people were exposed to accurate information regarding homosexuals. Leaders, too, tended to advocate sensitivity and diversity training especially in the earliest stages of an officer's career, although in police departments this was usually because a strong value was placed on officers' having the tools they needed to interact effectively with the gay and lesbian community. Heterosexual members of the rank and file of these organizations, however, were far more skeptical. When training was not perceived as being directly related to performing their job, they tended to resent the need to sit through discussions of lifestyles that they perceived as immoral or in which they had little interest. To their way of thinking, sensitivity training designed to facilitate the integration of gays into their forces was the very kind of coddling that signaled special class status and all the deleterious consequences that accompanied it. This was especially the case when such training took place in departments in which resources were clearly constrained, that is, where people were being laid off, benefits were being threatened, promotional opportunities were shrinking, and equipment was not being replaced because of budget shortfalls.

The Implementation Process

This study's attempt to understand how domestic police and fire departments implemented policies that allow acknowledged homosexuals to serve produced a number of insights into factors that influence the implementation process in both positive and negative ways. Most of these observations were articulated repeatedly by individuals across the variety of departments included in this study. A smaller number are based on analyses of the voluminous data collected as part of this endeavor. This section briefly highlights factors that facilitate and hinder the implementation *process,* and about how the implementation process itself tends to unfold. A more expansive discussion of these points can be found in Koegel (1993).

A first implementation lesson expressed by those interviewed in these police and fire departments was that nondiscrimination policies were most readily implemented when they were simple, clear, consistent, and forcefully stated. It was considered equally, if not more, important that they be *enforced* consistently, so that leadership at all levels was saying the same thing and practice matched the letter and spirit of formal policy. The departments included in this study were far more successful at accomplishing the former than the latter.

A second lesson expressed by the police officers and firefighters—particularly those in leadership—was that aggressive attempts to alter attitudes were foolhardy. Targeting *behavior,* they reported, was the appropriate approach. It was unreasonable, in other words, to expect members to give up strongly held and deeply entrenched beliefs overnight. It was not unreasonable, however, to insist that they keep those beliefs from interfering with their adherence to workplace expectations of behavior.

This suggested that policies of coexistence need not demand acceptance of homosexuals or homosexuality. The emphasis should be on controlling behavior through the enforcement of clear codes of conduct that were applicable to all servicemembers. Moreover, it was suggested that codes built on general principles of fairness, respect, honor, decorum, and the need to avoid the creation of hostile environments were far more practical and effective than those that tried to spell out every conceivable situation an officer might face.

A third lesson pertained to leadership. Leadership at all levels was unilaterally recognized as being critical to the successful implementation of controversial and potentially unpopular policies. The evidence was clear that strong leaders could push a department in one direction or another. The evidence also suggested how essential it was that commitment to the policy be internalized down the chain of command. It is worth noting that whereas the importance of leadership was unilaterally subscribed to, and whereas the emphasis at all levels of these organizations on following orders was consistently cited as facilitating policy implementation, none of the departments had achieved a real and consistent commitment to a policy of nondiscrimination across all ranks of leadership. Variability in the extent to which managers communicated and enforced messages sent down from the top—and, indeed, in the messages they were receiving from the top—was apparent in each.

Yet another implementation lesson expressed by many leaders and rank and file members suggested that when the solutions to problems related to integrating new groups into the work force either provide special privileges or inadvertently confer special class status, resentment and a host of more troubling problems on the part of the majority group may ensue (as was evident in the earlier discussion of the concerns of heterosexual members). This is not to say that special class protections are completely unwarranted. Out-groups are invariably at a significant disadvantage as they enter traditional organizations and may need assistance as these organizations adapt to their inclusion. But it was the opinion of many leaders and rank and file members that such assistance will be received far more gracefully if it is perceived as benefiting *all* members of the force. Fire departments, for instance, learned that there was far less resentment against women when firehouses were renovated to allow more privacy for *all* individuals than when private offices or common rooms were commandeered for use as bedrooms.

Similarly, the issue of training was viewed by these departments as having both advantages and disadvantages. Accurate information on who homosexuals are, how they come to be that way, and how they lead their lives was cited by many members of these departments, particularly leaders and homosexual members of the rank and file, as a potentially powerful tool in combating the stereotypic views held by many police officers and firefighters, especially if conducted by someone—preferably gay or lesbian—who has earned their respect in the workplace and knows what it means to do the work of the organization. But as described earlier, the responses of heterosexual members of the rank and file suggest that training can also draw ridicule and breed resentment, especially if it is seen as not being relevant to one's mission. Consequently, sensitivity training cannot unilaterally be viewed as positive and may even be inconsistent with the clearly articulated principle that as long as people adhere to behavioral guidelines, what they think is their own business. On the other hand, providing training to leaders on how best to implement a policy was always seen as being appropriate. Although good leadership usually prevails even in the absence of training, managers in these departments reported that the provision of support—helping leaders understand the policy, offering insights into how hypothetical situations might be handled, providing them with replies to the questions they might typically receive from those under their command—can improve their ability to effect positive change.

A last but extremely critical finding that emerged from the experiences of these police and fire departments pertains more to *how* the implementation process itself occurs. Regardless of when a formal policy of nondiscrimination toward homosexuals is enacted, change is not necessarily immediate. In reality, implementation proceeds at a pace that is particular to each institution and consistent with what it can absorb. Although the departments shared many things, each is situated in a different and ever-changing social climate, has its peculiar history and culture, draws upon slightly but significantly different pools of candidates for its work force, and has been influenced over time by very different sets of leaders. All of these combine to produce a unique level of tolerance in each department that constantly evolves. Data from this study suggest that neither the behavior of homosexuals in the workplace nor the aggressiveness with which the implementation of nondiscrimination policies occurs strays far from this tolerance level. This explains why so few lesbians and gay men publicly reveal their sexual orientation in these departments, particularly in fire departments. It also explains how a policy of nondiscrimination can be in place formally for significant periods of time, as was the case in several cities, but not result in any substantial departmental action toward implementation until years later.

This is not to say that actions never go beyond what might be perceived as tolerable by an organization. On rare occasions, homosexuals on the one hand and department leaders on the other may flirt with the threshold, and even advance beyond it. They invariably do so only slightly, however, provoking a mild and manageable reaction. In such situations, the effect of their actions often is to extend the boundaries of the threshold slightly. When they do so too aggressively, self-correcting mechanisms usually communicate their misjudgment and sustain the existing tolerance zone.

Implications of the Analogy for Implementing Policies of Nondiscrimination in the Military

This comprehensive examination of police and fire departments in six cities supports a number of critical findings and insights that are potentially relevant to the military's efforts to assess its policy toward homosexuals and to determine how the policy agreed upon can be implemented most effectively. This section reviews the most critical of these and elaborates on their potential relevance.

One set of findings has to do with the characteristics of those homosexuals drawn to paramilitary service and their behavior under a policy that permits them to disclose their sexual orientation. Homosexuals who join police and fire departments, first of all, do not fit stereotypes that are inconsistent with the image and mission of these organizations, and are attracted to police and fire work for the same reasons as are their heterosexual counterparts. Even more significantly, very few make their homosexuality public, particularly when their work environment is perceived as hostile to homosexuals and particularly in fire departments, in which work and living arrangements are more similar to those of the military. Finally, those who do make their homosexuality known are generally sensitive to the overall norms and customs of their organizations. They tend to behave in ways that neither shock nor offend, and subscribe to the organization's values on working problems out informally and within the ranks. What this suggests, of course, is that to the extent that the analogy of paramilitary organizations holds true for the military, the available evidence does not support concerns over who will enter the armed forces under a policy change, the extent to which homosexuals make their sexual orientation known, and the behavior of those who do. Indeed, the experience of police and fire departments suggests that the actual scope of the "problem" should a policy change be implemented in the military—that is, the actual number of known homosexuals—may be relatively small.

Another important set of findings pertains to the reactions of *heterosexual* officers under a policy of nondiscrimination based on sexual orientation. Although antihomosexual sentiment does not disappear after homosexuals acknowledge their sexual orientation, and although police and fire departments continue to be places in which hostility toward homosexuals is apparent, heterosexuals in these departments generally behave more mildly toward homosexuals than stated attitudes toward gays would predict. It is perhaps because of this tendency that homosexual officers usually perceive the consequences of acknowledging their sexual orientation to their departments as being manageable. In what may be seen as paradoxical, the prevailing rhetoric in these departments remains hostile to homosexuals—hostile enough to dissuade most individuals from disclosing their sexual orientation—but that rhetoric only rarely translates into dire consequences for those officers who do make their homosexuality publicly known. By extension, it may be a mistake to assume that the antihomosexual sentiment expressed within the military can necessarily be used as a valid representation of how heterosexuals will actually behave under a policy that allows gay men and lesbians to disclose their sexual orientation.

This is not to say that heterosexual officers do not express concerns that require the attention of those responsible for implementing policy. AIDS is a serious concern of heterosexual police and firefighters and not one that is quickly alleviated by education. The fear that homosexuals will receive special class protections is even more pronounced, however. The experience of police officers and firefighters suggests a need to protect homosexuals from harassment without conferring on them privileges that do not exist for majority groups.

As for implementation, lessons stemming from the experience of police and fire departments suggest the importance of policies that are unambiguous, consistently delivered, and uniformly enforced. The leaders of these departments emphasized that such policies need not be viewed as an endorsement of lifestyle or a statement about what is moral. When the primary emphasis in implementing policy is placed on changing behavior, not attitudes, members of a force can view homosexuality in any way they choose as long as their behavior is consistent with organizational codes of personal conduct. Such codes, the experience of police and fire departments suggests, should be written generically to restrict harassment and the creation of hostile environments vis-à-vis any force member to avoid the perception of special treatment. Training efforts that provide leaders with the information and skills they need to implement policy can facilitate the explication and enforcement of policy, as can the overriding value on discipline that is apparent in organizations such as the military. Sensitivity training for rank and file members of a force, however, may have mixed effects when it is not viewed as being explicitly related to performing one's job effectively.

A final finding regarding the implementation process speaks very clearly to the issue of whether a staged approach to implementing a policy change is necessary. The experience of domestic police and fire departments suggests that the implementation process with reference to policies of nondiscrimination against homosexuals is self-regulating, and that actual change occurs over long periods of time. In practice, homosexuals behave in ways that cluster around a zone of tolerance that may be unique to each organization and to settings within that organization. Moreover, the aggressiveness with which a nondiscrimination policy is pursued at an organizational level is similarly sensitive to organizational readiness for a change.

What this suggests is that policy actions calculated to slow the implementation process to allow actions to remain consistent with an organization's readiness for change are probably unnecessary. If the experience of police and fire departments is repeated in the military under a similar policy change, a stepwise implementation process and an overall conservative and measured reaction on the part of gay and lesbian officers can be expected to occur naturally over time. Change will happen, but rarely if ever will it move from point A to point Z regardless of whether stated policy, for the sake of simplicity and accuracy of intention, suggests that this is where it should go. Rather, it will take place in a more linear and staged fashion, with behaviors clustering around a readiness or tolerance threshold that constantly and inevitably adjusts itself. When attempts to codify "firebreaks" make the policy message more confusing, they may ultimately increase the difficulty of implementing a policy.

Conclusion

It cannot be predicted with certainty that a policy change within the military similar to the ones experienced by these police and fire departments will result in identical consequences, or that every lesson learned from these organizations can be applied directly to the armed forces. Consequently, this exercise has not "proved" anything. Even so, most of the insights that emerged from this examination of police and fire departments are not compromised by overwhelming threats to the analogy between public safety and military organizations. To the extent that this is true, such insights may be able to inform efforts to plan and implement policies regarding homosexuals in the military.

Notes

1. In no sense can the samples of rank and file members of these departments, either heterosexual or homosexual, be considered probability samples. Although every effort was made to ensure that those selected were representative of their departments, neither the methods used nor the sample size allows statements regarding the actual prevalence of the attitudes and behaviors described in subsequent sections. When evidence seemed strong on a given point, language conveys this. Otherwise, qualifiers that suggest precise prevalence estimates are deliberately avoided.

2. In departments in which hostility toward homosexuals was particularly strong, it was reported that individuals suspected of homosexuality were frequently harassed. A heterosexual man who had been subjected to persistent harassment because of such suspicions was one of several litigants in a recently settled lawsuit against one of the police departments in this study.

References

Alex, N. (1969). *Black in blue*. New York: Appleton-Century-Crofts.

Burke, M. (1993). *Coming out of the blue: British police officers talk about their lives in "the job" as gays, lesbians and bisexuals*. London: Cassell.

Burke, M. (1994). Homosexuality as deviance: The case of the gay police officer. *British Journal of Criminology, 34,* 192–203.

Craft, J. A. (1971). *Negroes in large municipal fire departments*. Lafayette, IN: Herman C. Krannert Graduate School of Industrial Administration, Purdue University.

Goldsmith, J., & Goldsmith, S. (Eds.). (1974). *The police community: Dimensions of an occupational subculture*. Pacific Palisades, CA: Palisades.

Heidensohn, F. (1992). *Women in control? The role of women in law enforcement.* Oxford: Clarendon Press.

Koegel, P. (1993). Analogous experience of domestic police and fire departments. In National Defense Research Institute, *Sexual orientation and U.S. military personnel policy: Options and assessment* (pp. 106–157). Santa Monica, CA: RAND.

Krueger, R. A. (1989). *Focus groups: A practical guide for applied research.* Newbury Park, CA: Sage.

Leinen, S. H. (1993). *Gay cops.* New Brunswick, NJ: Rutgers University Press.

Martin, S. E. (1980). *Breaking and entering: Policewomen on patrol.* Berkeley, CA: University of California Press.

Neiderhoffer, A., & Blumberg, A. S. (Eds.) (1970). *The ambivalent force: Perspectives on the police.* Waltham, MA: Ginn and Company.

National Institute of Mental Health. (1970). *Functions of the police in modern society: A review of background factors, current practices and possible role models.* Rockville, MD: Author.

Parachini, A. (1994). *Of the community and for the community: Race and gender integration in southern California police and fire departments.* Los Angeles: The Union.

Stewart, D. W., & Shamdasani, P. N. (1991). *Focus groups: Theory and practice.* Newbury Park, CA: Sage.

PART THREE

Cohesion, Privacy, and Attitudes

Sexual Orientation and Military Cohesion:
A Critical Review of the Evidence

Robert J. MacCoun

In early 1993, President Clinton directed the secretary of defense to draft an executive order that would end discrimination on the basis of sexual orientation in the military "in a manner that is practical, realistic, and consistent with the high standards of combat effectiveness and unit cohesion our Armed Forces must maintain."[1] This concern with military cohesion is not new. Cohesion has probably never been uniformly high (e.g., Griffith, 1988; Henderson, 1985; Manning and Ingraham, 1983; Scull, 1990; Siebold and Kelly, 1988), and the military intervenes whenever a unit becomes seriously dysfunctional for any reason. Because of this long-standing concern, a sizeable research literature exists on unit cohesion and its correlates. The present chapter critically reviews the cohesion literature and assesses its implications for the policy debate about sexual orientation and U.S. military personnel policy (see also MacCoun, 1993).

The analysis is premised upon three assumptions that appeared to be widely shared by both sides of the policy debate circa 1993–1995. First,

This essay is adapted from chapter 10 of the 1993 RAND National Defense Research Institute monograph, *Sexual orientation and U.S. military personnel policy,* the final report of a study commissioned by the office of the secretary, Department of Defense. I am grateful to Bernie Rostker, Scott Harris, John Winkler, Jim Kahan, and the rest of the RAND project team for their suggestions; to Bryan Hallmark, Susan Hosek, and Bruce Orvis for critical reviewing of the 1993 chapter; to Greg Herek for editorial suggestions; and to the many active and retired military personnel and cohesion experts who generously participated in interviews in 1993.

homosexuals are inherently no less capable of performing military tasks than are heterosexuals. Second, homosexuals already serve in the military, and always have, but most either have not openly acknowledged their status or have acknowledged it only to some colleagues. And third, if allowed to serve, homosexuals in the military would be held to standards of conduct, appearance, demeanor, and performance at least as stringent as the standards for heterosexuals. Thus, concerns about cohesion pertain to acknowledged homosexual status, not sexual orientation per se, and to how an individual's acknowledged homosexuality would affect the group. Given these assumptions, the central question of the chapter is: What effect will the presence of acknowledged homosexuals have on the cohesion and performance of a given military unit?

Cohesion and Its Effects on Performance

What Is Cohesion?

Many early academic definitions of cohesion emphasized the quality of the social relationships among group members; for example, "that group property which is inferred from the number and strength of mutual positive attitudes among the members of a group" (Lott and Lott, 1965, p. 259). Military definitions, in contrast, tend to define cohesion in the context of the combat mission; for example, "cohesion exists in a unit when the primary day-to-day goals[2] of the individual soldier, of the small group with which he identifies, and of unit leaders, are congruent—with each giving his primary loyalty to the group so that it trains and fights as a unit with all members willing to risk death and achieve a common objective" (Henderson, 1985, p. 4).

Investigators have developed measures of cohesion with adequate reliability (e.g., Carron, Widmeyer, and Brawley, 1985; Siebold and Kelly, 1988). A more persistent problem involves construct validity. Researchers frequently have failed to distinguish among a variety of concepts that are often listed as aspects of cohesion. Cohesion itself has been equated with such diverse concepts as motivation, satisfaction, mutual friendship, caring, interpersonal attraction, shared goals, teamwork, coordination, group pride, group prestige, and group status. This imprecision produced considerable confusion about the concept and its effects on performance until the 1980s, when there was a renewed recognition of early calls for a distinction between different types of cohesion (e.g., Davis, 1969; Shaw, 1976; Steiner, 1972).

Although a variety of labels have been used, the most common distinction is between *social cohesion* and *task cohesion* (e.g., Carron, Widmeyer, and Brawley, 1985; Davis, 1969; Griffith, 1988; Mudrack, 1989; Mullen and Copper, 1994; Siebold and Kelly, 1988; Tziner and Vardi, 1982; Zaccaro and McCoy, 1988).[3] Social cohesion refers to the nature and quality of the emotional bonds of friendship, liking, caring, and closeness among group members. A group is socially cohesive to the extent that its members like each other, prefer to spend their social time together, enjoy each other's company, and feel emotionally close to one another. Task cohesion refers to the shared commitment among members to achieving a goal that requires the collective efforts of the group. A group with high task cohesion is composed of members who share a common goal and who are motivated to coordinate their efforts as a team to achieve that goal. This general distinction is supported by both experimental and correlational evidence (e.g., Anthony et al., 1993; Carron, Widmeyer, and Brawley, 1985; Griffith, 1988; Mullen and Copper, 1994; Mullen et al., 1994; Siebold and Kelly, 1988; Zaccaro and McCoy, 1988).

What Effect Does Cohesion Have on Unit Performance?

Many reviewers have struggled to make sense of the conflicting results across studies of the cohesion-performance relationship, in part because the relevance of the social-task distinction was not fully appreciated (e.g., Lott and Lott, 1965; Mudrack, 1989; Shaw, 1976). Recent applications of meta-analytic methods have provided greater clarity by statistically aggregating results across independent studies. Using overlapping collections of studies, each meta-analysis indicates that, overall, there is a modest positive relationship between cohesion and performance. Oliver's (1990b) meta-analysis at the Army Research Institute for the Behavioral and Social Sciences (ARI) reported an average correlation of $r = .32$. Evans and Dion's (1991) meta-analysis reported an average correlation of $r = .36$. The most complete meta-analysis was conducted by Mullen and Copper (1994) under contract to ARI. They identified forty-nine studies containing sixty-six separate estimates of the cohesion-performance link, and found an average correlation of $r = .25$.

Of course, a reliable correlation between cohesion and performance could simply reflect the causal influence of performance on cohesion (Oliver, 1990a). Using meta-analytic and adjusted cross-lagged panel analysis techniques, Mullen and Copper (1994) concluded that the effect of success on cohesion was actually larger than the effect of cohesion on performance. This conclusion is bolstered by evidence from experimental manipulations of cohesion and performance (see Lott and Lott, 1965).

Moderating factors. The Mullen and Copper meta-analysis provided a detailed examination of a number of variables that appear to moderate the cohesion-performance relationship. The association is strongest for sports teams ($r = .54$, $n = 8$ tests), substantially weaker for military units ($r = .23$, $n = 10$ tests) and other real work groups ($r = .20$, $n = 13$ tests), and weakest for artificial groups ($r = .16$, $n = 12$ tests).

Janis (1983) suggested that "the duality of cohesiveness may explain some of the inconsistencies in research results on group effectiveness" (p. 248). This argument is supported by the Mullen and Copper (1994) meta-analysis,[4] which indicated that only task cohesion was independently associated with group performance; social cohesion and group pride were not correlated with performance after statistically controlling for task cohesion. This suggests that it is *task cohesion,* not social cohesion or group pride, that drives group performance. This conclusion is consistent with the results of hundreds of studies in the industrial-organizational psychology literature on the crucial role of goal setting for productivity (see Locke and Latham, 1990).

Deleterious effects. The lack of an independent effect of social cohesion in experimental studies, and the negative effect of social cohesion among correlational studies, may seem counterintuitive. But both military (e.g., Driskell, Hogan, and Salas, 1987; Tziner and Vardi, 1982; see MacCoun, 1993, for more examples) and nonmilitary (e.g., Davis, 1969; Janis, 1983; Lott and Lott, 1965) reviews have noted that social cohesion has complex and sometimes deleterious effects on group performance. Note that the claim that high social cohesion sometimes undermines performance should not be taken to imply that low social cohesion is actually desirable; it is not. Janis (1983) proposed that "for most groups, optimal functioning in decision-making tasks may prove to be at a moderate level of cohesiveness" (p. 248).

High social cohesion can result in excessive socializing, which interferes with task performance (see Davis, 1969; Lott and Lott, 1965; Zaccaro and McCoy, 1988). Davis (1969, p. 79) noted that the "pleasure from interaction itself, in cohesive groups, sometimes exceeds the task-specific motivation, and greater energy is devoted to interpersonal relations than to overcoming the task obstacles. Hence performance suffers." According to Janis (1983), the probability of *groupthink*—a condition leading to defective group decision processes—is stronger "when high cohesiveness is based primarily on the rewards of being in a pleasant 'clubby' atmosphere or of gaining prestige from being a member of an elite group than when it is based primarily on the opportunity to function competently on work tasks with effective co-workers" (p. 247). A recent meta-analysis (Mullen et al., 1994) linked groupthink to social cohesion, but not task cohesion. In the field of organizational behavior, a common example of the deleterious effects of cohesion is *rate-busting*—an agreement among workers, either tacitly or explicitly, to maintain low levels of performance (e.g., Bass, 1981; Janis, 1983). In the military context, there are many more serious examples of these deleterious consequences, including drug use, insubordination, and mutiny (Ingraham, 1984).

Factors Influencing Social and Task Cohesion

A sizable research literature describes the factors that promote cohesion (reviewed in MacCoun, 1993). Unfortunately, many studies focused exclusively on social cohesion, or failed to distinguish social from task cohesion. Consequently, the antecedents of social cohesion are somewhat better understood than those of task cohesion.

Propinquity and Group Membership
Ingraham (1984) argued that in Army barracks, "by far the most potent determinant of social choice [of friends] was the company of assignment" (p. 58). This conclusion is amply supported by the research literatures on the link between propinquity and friendship (e.g., Lott and Lott, 1965), and the effects of randomly assigning individuals to ad hoc groups (Tajfel and Turner, 1979; Wilder, 1986). Thus, the simple fact that individuals are assigned to a unit together predisposes them to social cohesion.

Leadership
Military analysts have identified the quality of leadership as a key factor in determining whether units are cohesive (Henderson, 1985; Manning and Ingraham, 1983; Siebold and Kelly, 1988). According to Scull (1990), "cohesion among soldiers remains primarily the by-product of good leadership combined with important, fulfilling work" (p. 24). This

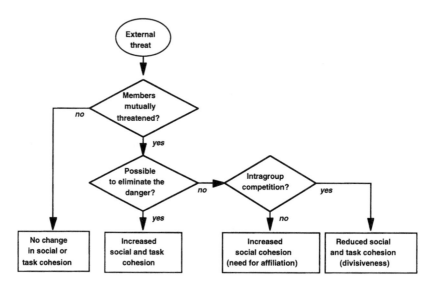

Figure 8.1. Effects of external threats on social and task cohesion.

claim is supported in nonmilitary organizations as well (e.g., Bass, 1981; Hollander, 1985).

Group Size

Group cohesion is inversely related to group size (Mullen and Copper, 1994; Siebold and Kelly, 1988; Steiner, 1972). Siebold and Kelly (1988) suggested that the platoon was the optimal size for measuring cohesion. Savage and Gabriel (1976) argued that "in conflict, the unit of cohesion tends to be the squad" (p. 364). That cohesion declines with group size explains why larger groups have weaker cohesion-performance correlations (Mullen and Copper, 1994).

Success Experiences

As reviewed above, successful performance experiences promote cohesion. Indeed, the effect of performance on cohesion appears to be stronger than the effect of cohesion on performance.

Shared Threat

Numerous studies suggest that under certain conditions, external threats can enhance cohesion (e.g., Dion, 1979; Schachter, 1959; Sherif et al., 1961; Stein, 1976). Figure 8.1 outlines these conditions. First, the group members must be mutually threatened. If so, their response depends on whether they perceive the possibility of a collective response that will eliminate the danger. If so, both social and task cohesion are likely to be enhanced (Johnson and Johnson, 1989; Sherif et al., 1961; Stephan, 1985). Sherif et al.'s classic studies demonstrated that in the face of a superordinate threat and goal, even hostile groups can merge together to form a cohesive whole. For example, during World War II, Brophy (1945–1946) found that white seamen's prejudice against African American seamen was inversely associated with shared exposure to enemy fire. Brophy concluded that "it would appear that many of our respondents could not afford the luxury of an anti-Negro prejudice while at sea" (p. 466).

But even when no collective instrumental response is available, anxiety appears to promote affiliation or social cohesion (e.g., Schachter, 1959), except perhaps when scarcity creates a conflict between personal and group interests (Hamblin, 1958, cited in Stein, 1976). This social cohesion may sometimes be temporary. Moskos (quoted in Marlowe, 1979) has suggested that earlier scholars failed to appreciate the extent to which the bonding in combat situations was "instrumental and self-serving" (p. 52), a temporary and situational adaptation to danger. Thus, much of what appears to be social cohesion on the battlefield may have more to do with task cohesion or tacit psychological contracts ("I'll cover you if you'll cover me") than with the intrinsic likability of one's comrades.

Similarity/Homogeneity

The positive association between interpersonal liking and similarity of attitudes, interests, and values is well established (e.g., Lott and Lott, 1965). A meta-analysis of seventeen studies comprising twenty-five separate estimates (Anthony et al., 1993) yielded an average similarity-cohesion correlation $r = .24$. But the effect was significantly weaker in enduring groups—e.g., military units, sports teams, work groups—than in temporary, artificially created laboratory groups, and it decreased with group size and with the ratio of males to females in the group. Similarity of attitudes and values appeared to be unrelated to *task* cohesion in either type of group. An ARI study (Siebold and Lindsay, 1994) indicated that racial/ethnic heterogeneity similarly lacks any reliable relationship to cohesion in military units. Thus, similarity does not appear to influence task cohesion, which is the type of cohesion that influences group performance. This is consistent with other research suggesting that group homogeneity has no net effect on productivity (e.g., Steiner, 1972; Shaw, 1976).

Addressing the Effects of Lifting the Ban

Those who oppose allowing open acknowledgment of homosexual orientation in the military cite concerns about how the presence of acknowledged homosexuals would affect unit cohesion and performance. In this section, I address four questions about a change in policy. First, how many units would have acknowledged homosexuals as members? Second, how might the presence of an acknowledged homosexual influence task and social cohesion? Third, would heterosexual personnel's negative attitudes toward homosexuality be expressed behaviorally? And fourth, would heterosexuals obey an acknowledged homosexual leader?[5]

Because no systematic empirical research has been conducted on the effect of acknowledged homosexuals on unit cohesion or unit performance, the conclusions in this section are based on established social psychological theory and the general empirical findings of the cohesion literature.

Will Many Units Have Acknowledged Homosexuals as Members?
Almost all experts agree that the prevalence of homosexual behavior in the adult population falls somewhere in the 1 percent to 10 percent range (see chapter 2). But many of those who engage in homosexual behavior also engage in heterosexual behavior and do not consider themselves to be homosexual. Consequently, the prevalence of individuals with a homosexual self-identity—whether overt or covert—is probably nearer to the low end than the high end of that range. As noted in the overview of this chapter, however, many of the concerns raised in the policy debate involve not the prevalence of homosexuality in the military, but the prevalence of individuals who *openly acknowledge* a homosexual orientation. In reality, the openness of one's sexual orientation is not a dichotomous variable, but a continuous variable. For example, some homosexuals might be open only to close friends, but not to mere acquaintances (see chapter 10). With respect to the hypothesized threat posed to military cohesion, an appropriate operational definition of openness is the extent to which someone's homosexual orientation is acknowledged by the individual, and known by a majority of the individual's colleagues and by supervisors.

Using this definition, the experiences of domestic paramilitary institutions that have adopted nondiscrimination policies are relevant. As shown in the table below, police and fire departments visited by the RAND team reported that between 0 and 0.51 percent of their total membership consisted of acknowledged homosexuals, with a mean prevalence of 0.14 percent and a median prevalence of 0.07 percent (see also chapter 7). Coming out (i.e., disclosing one's homosexual orientation), even in an officially nondiscriminatory atmosphere, is a risky choice. Homosexuals can face hostility from colleagues, unequal treatment from supervisors, and even physical violence. Gay and lesbian members of these organizations appear to be generally unwilling to acknowledge their sexual orientation unless the local climate is tolerant (see chapter 10).

This suggests that *acknowledged* homosexuals would likely be quite rare in the military, even if all restrictions on service by homosexuals were removed. Recall that group cohesion is mostly relevant at the level of platoons (sixteen to forty members) and smaller units, such as five-

Estimated Prevalence of Acknowledged Homosexuals in Domestic Paramilitary Institutions Visited by RAND

Institution	Location	Year Policy Changed	Total Force Size	Number of Acknowledged Homosexuals	Prevalence of Acknowledged Homosexuals (percentage)
Police	Chicago	1988	12,209	7	0.06
	Houston	1990–91	4,100	0	0.00
	Los Angeles	1979	7,700	7	0.09
	New York[a,b]	1979	28,000	approx. 100	0.36
	San Diego[b]	1990	1,300	4 or 5	0.25
	Seattle[c]	1980	1,300	2	0.15
Fire	Chicago	1988	4,700	0	0.00
	Houston	1990–91	2,900	0	0.00
	Los Angeles	1979	3,200	0	0.00
	New York	1979	11,300	0	0.00
	San Diego[d]	1990	845	1	0.12
	Seattle[d]	1980	975	5	0.51
Mean					0.14
Median					0.07

[a] Acknowledged homosexual officers are actively recruited for community policing in heavily homosexual neighborhoods.

[b] We were unable to get a precise count of acknowledged homosexuals.

[c] After our visit, the *Seattle Times* reported the resignation of an acknowledged homosexual in the Seattle Police Department (Shatzkin, 1993).

[d] The only acknowledged homosexual firefighters in the cities we visited were lesbians.

person teams or crews. If the current policy were changed and if the prevalence of open homosexuals in the military were to match the mean prevalence in the paramilitary institutions studied by the RAND team, fewer than 6 percent of forty-person platoons and fewer than 1 percent of five-person crews and teams would be expected to have an open homosexual. And just a small fraction of 1 percent of platoons would have two or more open homosexuals. If the presence of open homosexuals were to be clustered rather than randomly distributed, for any given aggregate prevalence rate, even fewer units would have an open homosexual.

Will the Presence of Acknowledged Homosexuals Influence Cohesion?

Although there is no direct scientific evidence, the established principles of cohesion suggest that if the presence of acknowledged homosexuals has an effect, it is most likely to involve social cohesion rather than task cohesion. As explained above, similarity of social attitudes and beliefs is not associated with task cohesion, although it is sometimes associated with social cohesion. Task cohesion involves similarity, but of a different sort; it is found when individuals share a commitment to the group's purpose and objectives. There seems little reason to expect acknowledged homosexuality to influence this commitment, at least not directly. The values of homosexuals in the military have not been systematically compared to those of heterosexual personnel. But historical anecdotes and RAND's interviews with closeted gay and lesbian military personnel suggest that homosexuals who serve in the military are committed to the military's core values, which Henderson (1990) lists as "fighting skill, professional teamwork, physical stamina, self-discipline, duty (selfless service), and loyalty to unit" (p. 108). Commitment to these values seems particularly likely, given that homosexuals in the military are a self-selected group and enlist despite numerous obstacles and personal and professional risks.

Although any reduction in cohesion is likely to involve social cohesion, as explained previously, it is task cohesion rather than social cohesion that has a direct influence on performance. This indicates that it is not always necessary for coworkers to like each other, or desire to socialize together, to perform effectively as a team. According to Steiner (1972), "work groups sometimes persist in the face of adversity even though members have little affection for one another" (p. 161). There are many examples of this phenomenon in the sports literature, including the 1973–1975 Oakland A's and the 1977–1978 New York Yankees. Aronson (1976) described how black and white coal miners in West Virginia "developed a pattern of living that consisted of total and complete integration while they were under the ground, and total and complete segregation while they were above ground" (p. 193). Many military observers (e.g., Ingraham, 1984; see also Schlossman et al., 1993) have noted that African American and white soldiers tended to socialize separately despite working together effectively.

Under some conditions, however, a reduction in social cohesion may bring about a reduction in task cohesion. For certain types of tasks, some minimal level of social cohesion might be necessary for the group to accomplish its task (Driskell, Hogan, and Salas, 1987; Janis, 1983; Zaccaro and McCoy, 1988). One might expect this to be less of a concern in *additive* tasks (in which the group's performance is the sum of individual performances), and more of a concern in *disjunctive* and *conjunctive* tasks (in which the group's performance is determined by the most able member or the "weakest link," respectively; Steiner, 1972; Zaccaro and McCoy, 1988). Mullen and Copper's (1994) meta-analysis did not support this prediction, however; it found no differences in the strength of the cohesion-performance effect for tasks with high versus low interactive requirements.

Nevertheless, one can imagine circumstances in which a group has so little social cohesion that task performance becomes impossible, with potentially disastrous consequences for the group. Thus, much may depend on how social cohesion is affected. Figure 8.2 presents four qualitative types of social cohesion in a five-person crew or team, in which individual E has revealed his or her homosexual orientation.[6]

In figure 8.2, model *a* depicts a group in which social cohesion has not been disrupted. Model *b* depicts the complete breakdown of social cohesion—a state of anarchy. This situation would prevail if E's acknowledgment of homosexuality actually affected the bonds of friendship among *heterosexuals* in the unit; for example, A likes C less because E is a homosexual. This scenario seems quite unlikely. Model *c* is somewhat more plausible. In this scenario, the crew is split into factions; members A, B, and C are hostile to the homosexual and to D, who befriends the homosexual. This is conceivable only if D is willing to sacrifice his relationships with the others in the process. Of course, D may also be a homosexual, though statistically this will be unlikely.

If there is any breakdown in social cohesion, model *d* describes the most likely scenario: the case of complete or partial ostracism. Because

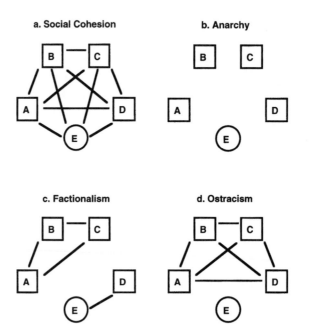

Figure 8.2. Alternate states of cohesion in five-person unit. (Note: Node E depicts an acknowledged homosexual; links depict positive bonds between individuals.)

ostracism provides others in the group with a common enemy, the strength of the bonds among the heterosexuals might actually increase, resulting in a net increase in social cohesion for the group as a whole. That is not to suggest that ostracism is in any way acceptable. When ostracism occurs—for any reason—the military actively intervenes through informal conflict resolution, reassignment, or disciplinary action. Whether complete ostracism occurs would probably depend partly on the performance and demeanor of the acknowledged gay man or lesbian, but mostly on whether the unit's other members would refuse to cooperate with each other to accomplish the group's mission. This latter issue is addressed below.

Will Negative Attitudes Toward Homosexuality Be Expressed Behaviorally?

Heterosexual military personnel's expression of negative attitudes toward homosexuality has raised concerns about how these personnel will behave if they find themselves working with an acknowledged homosexual.

Thus, various opponents of the ban have predicted that soldiers will re-fuse to work, bunk, or shower with homosexuals, and that there will be widespread outbreaks of violence against homosexuals. But there is little reason to believe that negative attitudes toward homosexuality are auto-matically translated into destructive behaviors. The effect of attitudes to-ward social groups on behavior is known to be indirect, complex, and, for most people, fairly weak (Ajzen and Fishbein, 1980; Eagly and Chaiken, 1993; LaPiere, 1934; Stephan, 1985).

At one time, behavioral researchers simply assumed that social atti-tudes were a major determinant of behavior. An early indication that this might not be the case was provided by LaPiere (1934). He traveled across the United States with a Chinese couple and found that of approximately two hundred and fifty hotels and restaurants, only one refused to serve the couple. LaPiere then surveyed the proprietors of these institutions to ask if their establishments accepted Chinese people; of the 128 replies he received, more than 90 percent said that they did not. Stephan (1985, p. 627) cited several replications of this finding involving similar discrepan-cies between expressed prejudice and actual behaviors toward African Americans. In light of these and other studies, many social scientists con-cluded that such attitudes have little or no association with behavior.

In recent decades, there has been considerable research on ways in which attitudes actually do influence behavior (see Ajzen and Fishbein, 1980; Eagly and Chaiken, 1993; Triandis, 1977). Figure 8.3 illustrates a number of important points about the relationship between attitudes toward subjects—in this case, attitudes toward homosexuals—and be-havior. First, the relationship between attitudes toward subjects or ob-jects and actual behaviors is quite indirect. A negative attitude toward homosexuality will influence behavior only via its influence on attitudes *toward acts;* for example, the attitude toward working with (or sleeping or showering next to) this homosexual. Moreover, attitudes toward ho-mosexuality are only partial determinants of attitudes toward these acts. Behaviors are also determined by their perceived consequences: the per-ceived benefits ("I'll avoid having to be around someone I don't like; others will know that I'm not homosexual;" etc.), and costs ("We won't get the job done; I'll interfere with the unit's mission; I may end up in an unpleasant confrontation with the homosexual person; I may have to endure disciplinary actions by my superiors").

Moreover, the attitude toward the act is itself only indirectly related to behavior through its influence on the intention to engage in the act,

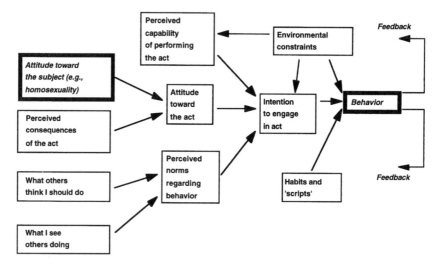

Figure 8.3. Indirect nature of the attitude-behavior link. (Note: Arrows depict causal relationships.)

which may be influenced by attitudes, but also by injunctive and descriptive social norms (Cialdini, Kallgren, and Reno, 1991). Injunctive norms refer to our beliefs about what we think others want us to do; for example, in deciding whether to refuse to work with a homosexual, I may anticipate the approval of my heterosexual buddies, but the disapproval of my supervisor. Descriptive norms refer to what we see others doing in similar situations; for example, if I see my heterosexual peers working with the homosexual soldier, I will be more inclined to work with him too.

Intentions are also influenced by the actor's perceived capability of performing the act (Bandura, 1982). In many situations, it may be quite difficult to refuse contact with a homosexual: "If I don't ride with this guy, how will I get there? If I refuse to sleep next to him, where will I sleep?" Finally, behavior itself is only partly intentional. Like intentions, behaviors are also constrained by the resources and opportunities afforded by the immediate environment. And our behaviors in many situations reflect well-learned habits or behavioral scripts that we engage in with little or no conscious reflection (Eagly and Chaiken, 1993; Triandis, 1977). Cooperating with coworkers is a fairly automatic, habitual response to work situations for most adults.

The principles depicted in Figure 8.3 help to explain why diffuse attitudes toward social groups have only weak effects on behavior. This is not to say that negative attitudes toward homosexuality will never be expressed behaviorally; history clearly suggests otherwise. But figure 8.3 indicates that many factors mitigate against serious behavioral expressions of anti-homosexual attitudes. The military has considerable influence over many of those mitigating factors—the consequences of the action, the injunctive and descriptive norms, the environmental constraints, habits and scripts—through its leadership, its regulations, its standard operating procedures, and its training and socialization process. If military leaders set and enforce clear standards for acceptable and unacceptable conduct, compliance is likely to be high. It will not be universal, however. Some individuals will test their leaders' resolve to enforce compliance. Leaders who display ambivalence about enforcement can probably anticipate further problems (see chapter 13).

Because of their compliance with regulations that conflict with their own attitudes, many individuals may experience cognitive dissonance (for a review, see Eagly and Chaiken, 1993). When individuals with negative attitudes toward homosexuality find themselves cooperating with acknowledged homosexuals, they may resolve this sense of dissonance in a number of different ways. They might behave in a passive-aggressive manner, engage in symbolic displays of heterosexual identity, use fear of punishment or a sense of duty to rationalize their actions, or change their attitudes toward homosexuals. Dissonance frequently produces attitude change, but it should be emphasized that the goal of compliance is to establish unit discipline, cohesion, and effectiveness. The military can demand cooperation and compliance with regulations without requiring a change in attitudes toward homosexuality.

Will Heterosexuals Obey an Openly Homosexual Leader?
Earlier, it was suggested that if social cohesion is adversely affected by the integration of open homosexuals, it is most likely to occur through a process of partial or complete ostracism. What if the ostracized homosexual is the group's leader? Will heterosexual soldiers respect an acknowledged homosexual, and comply with his or her orders? Relevant to this point, similar concerns were raised about African American military leaders before racial desegregation, but were not actually borne out by postintegration experience (see Schlossman et al., 1993).

Military leaders obviously benefit from being liked, but it may not be necessary to be liked to get the job done; indeed, Bass (1981) noted that perceived ability is a more important source of esteem than is popularity. This is consistent with Hollander's (1985) *idiosyncracy credit* model, which suggests that group members must demonstrate their competence and their loyalty to the group before they can deviate from group norms. In cases in which the leader's sexual orientation is known before joining a unit, many subordinates will obey a homosexual leader simply because of a strong sense of duty. Ultimately, then, much may depend on the behavior of the next leader up the chain of command. If the homosexual leader is treated with respect from above, he or she is more likely to be treated with respect from below.

Conclusions

Although concerns about the potential effect of permitting homosexuals to serve in the military are not groundless, the likely problems are not insurmountable, and there is ample reason to believe that heterosexual and homosexual personnel can work together effectively. The presence of acknowledged homosexuals may reduce social cohesion in some units, but seems unlikely to undermine task cohesion. Research indicates that it is not necessary to like someone to work with him or her, so long as members share a commitment to the group's objectives. If there is a reduction in social cohesion, it will probably involve some degree of ostracism of the homosexual, rather than a complete breakdown of the unit. Whereas some heterosexuals might refuse to cooperate with known homosexuals, many factors will discourage this and promote teamwork: effective leadership; military norms, roles, regulations, and disciplinary options; and external threats and challenges.

Even if the current policy were to be changed, openly homosexual military personnel are likely to be rare, at least in the foreseeable future. Homosexuals in the military will be under enormous informal pressure to stay in the closet, even without any explicit requirement to do so. As a result, only a small minority of platoons and teams are likely to have acknowledged homosexuals as members. This low prevalence will help to limit the frequency of conflicts, but it will also limit opportunities for the kind of positive social interaction that helps to overcome stereotypes (see chapters 9 and 10; Stephan, 1985).

Because there are no systematic scientific investigations of the effects of acknowledged homosexual identity on the cohesion of military units, these conclusions are based on indirect inferences from existing theory and research on cohesion, small group behavior, and attitude-behavior relations. Nevertheless, these conclusions are supported by the experiences of foreign military and American police and fire departments, as well as the lessons of racial desegregation in the military (see chapters 5, 6, and 7; National Defense Research Institute, 1993). Although none of these situations provide perfect analogies, each provides two out of three key elements—the United States, the military, and sexual orientation—and is at least roughly analogous in its other elements. This convergence of findings across disparate research literatures and case studies is all too rare in the social and policy sciences, and should enhance our confidence in these conclusions.

Notes

1. Memorandum for the Secretary of Defense, Ending Discrimination on the Basis of Sexual Orientation in the Armed Forces, January 29, 1993.

2. As written, this is a good definition of *task cohesion* (defined later). But in his prepared statement to the Senate Armed Services Committee hearings (1993), Colonel Wm. Darryl Henderson substituted the phrase "primary *values and* day-to-day goals" (emphasis added). As explained below, this revision is not supported psychometrically, unless "values" refers to teamwork and endorsement of the unit's mission.

3. Mullen and Copper (1993) use the terms *interpersonal attraction* and *commitment to task*. Siebold and Kelly (1988a) use the terms *affective bonding* and *instrumental bonding*. Tziner and Vardi (1982) use the terms *socio-emotional cohesiveness* and *task-oriented (instrumental) cohesiveness*. Zaccaro and McCoy (1988) use the terms *interpersonal cohesiveness* and *task-based cohesiveness*. MacCoun (1993) lists additional examples. This proliferation of terms has added to the confusion in the literature; on the other hand, it indicates that several different research teams have more or less independently recognized the need for this distinction.

4. Specifically, for each correlational study, they coded the proportion of questionnaire items tapping interpersonal attraction, commitment to task, and group pride. For experimental studies, four judges each rated the manipulations of cohesion with respect to the three types of cohesion. Because these three dimensions of cohesion were correlated—positively for experimental studies; negatively for correlational studies—they computed residual measures of social cohesion, task cohesion, and group pride, partialling out their shared variance.

5. A fifth question—"Will contact with acknowledged homosexuals influence attitudes toward homosexuality?"—is addressed elsewhere (see chapters 9 and 10; MacCoun, 1993).

6. Social cohesion involves the pairwise bonds among these individuals.

Strictly speaking, there should be two directional bonds for each pair of individuals, but figure 8.2 depicts only one, for simplicity. Similarly, in reality, these bonds vary continuously in strength, but figure 8.2 treats them dichotomously for simplicity. It assumes that if either individual rejects the other, the pairwise bond is broken; this is a pessimistic assumption that provides an upper bound on the loss of cohesion.

References

Ajzen, I., & Fishbein, M. (1980). *Understanding attitudes and predicting social behavior.* Englewood Cliffs, NJ: Prentice-Hall.

Anthony, T., Copper, C., Dovidio, J. F., Driskell, J. E., Mullen, B., & Salas, E. (1993). *The effect of group member similarity on cohesiveness: Do birds of a feather really flock together?* Unpublished manuscript, Department of Psychology, Syracuse University.

Aronson, E. (1976). *The social animal* (2d ed.). San Francisco: W.H. Freeman.

Bandura, A. (1982). Self-efficacy mechanism in human agency. *Psychological Review, 84,* 191–215.

Bass, B. M. (1981). *Stogdill's handbook of leadership* (revised and expanded edition). New York: Free Press.

Brophy, I. N. (1945–1946). The luxury of anti-Negro prejudice. *Public Opinion Quarterly, 9,* 456–466.

Carron, A. V., Widmeyer, W. N., & Brawley, L. R. (1985). The development of an instrument to assess cohesion in sport teams: The Group Environment Questionnaire. *Journal of Sport Psychology, 7,* 244–266.

Cialdini, R. B., Kallgren, C. A., & Reno, R. R. (1991). A focus theory of normative conduct: A theoretical refinement and reevaluation of the role of norms in human behavior. In M. Zanna (Ed.), *Advances in experimental social psychology* (Vol. 24, pp. 201–234). New York: Academic Press.

Davis, J. H. (1969). *Group performance.* Reading, MA: Addison-Wesley.

Dion, K. L. (1979). Intergroup conflict and intragroup cohesiveness. In W. G. Austin & S. Worchel (Eds.), *The social psychology of intergroup relations* (pp. 211–224). Monterey, CA: Brooks/Cole.

Driskell, J. E., Hogan, R., & Salas, E. (1987). Personality and group performance. In C. Hendrick (Ed.), *Group processes and intergroup relations.* Newbury Park, CA: Sage.

Eagly, A. H., & Chaiken, S. (1993). *The psychology of attitudes.* Fort Worth, TX: Harcourt Brace Jovanovich.

Evans, N. J., & Dion, K. L. (1991). Group cohesion and performance: A meta-analysis. *Small Group Research, 22,* 175–186.

Griffith, J. (1988). Measurement of group cohesion in U.S. Army units. *Basic and Applied Social Psychology, 9,* 149–171.

Hamblin, R. L. (1958). Group integration during a crisis. *Human Relations, 11* (1), 67–76.

Harry, J. (1984). Homosexual men and women who served their country. *Journal of Homosexuality, 10,* 117–125.

Henderson, W. D. (1985). *Cohesion: The human element.* Washington, DC: National Defense University Press.

———. (1990). *The hollow army.* New York: Greenwood Press.

———. (1993, March 31). *Prepared statement for Senate Armed Services Committee.* Washington, DC: U.S. Government Printing Office.

Hollander, E. P. (1985). Leadership and power. In G. Lindzey & E. Aronson (Eds.), *The handbook of social psychology* (3rd ed., Vol. 2, pp. 485–538). New York: Random House.

Ingraham, L. H. (1984). *The boys in the barracks: Observations on American military life.* Philadelphia: ISHI.

Janis, I. L. (1983). *Groupthink: Psychological studies of policy decisions and fiascoes* (2d ed.). Boston: Houghton Mifflin.

Johnson, D. W., & Johnson, R. T. (1989). *Cooperation and competition: Theory and research.* Edina, MN: Interaction Book Company.

LaPiere, R. T. (1934). Attitudes vs. actions. *Social Forces, 13,* 230–237.

Locke, E. A., & Latham, G. P. (1990). Work motivation and satisfaction: Light at the end of the tunnel. *Psychological Science, 1,* 240–246.

Lott, A. J., & Lott, B. E. (1965). Group cohesiveness as interpersonal attraction: A review of relationships with antecedent and consequent variables. *Psychological Bulletin, 64,* 259–309.

MacCoun, R. J. (1993). What is known about unit cohesion and performance. In National Defense Research Institute, *Sexual orientation and U.S. military personnel policy: Options and assessment* (pp. 283–321). Santa Monica, CA: RAND.

Manning, F. J., & Ingraham, L. H. (1983). *An investigation into the value of unit cohesion in peacetime.* Washington, DC: Walter Reed Army Institute of Research.

Marlowe, D. H. (1979). *Cohesion, anticipated breakdown, and endurance in battle: Considerations for severe and high intensity combat.* Unpublished manuscript, Walter Reed Army Institute of Research, Washington, DC.

Mudrack, P. E. (1989). Defining group cohesiveness: A legacy of confusion? *Small Group Behavior, 20,* 37–49.

Mullen, B., Anthony, T., Salas, E., & Driskell, J. E. (1994). Group cohesiveness and quality of decision making: An integration of tests of the groupthink hypothesis. *Small Group Research, 25,* 189–204.

Mullen, B., & Copper, C. (1994). The relation between group cohesiveness and performance: An integration. *Psychological Bulletin, 115,* 210–227.

National Defense Research Institute (1993). *Sexual orientation and U.S. military personnel policy: Options and assessment.* Santa Monica, CA: RAND.

Oliver, L. W. (1990a). *Cohesion research: Conceptual and methodological issues.* Alexandria, VA: U.S. Army Research Institute for the Behavioral and Social Sciences.

Oliver, L. W. (1990b). *The relationship of group cohesion to group performance:*

A meta-analysis and critique of the literature. Alexandria, VA: U.S. Army Research Institute for the Behavioral and Social Sciences.

Savage, P. L., & Gabriel, R. A. (1976). Cohesion and disintegration in the American Army: An alternative perspective. *Armed Forces and Society, 2,* 340–376.

Schachter, S. (1959). *The psychology of affiliation.* Stanford, CA: Stanford University Press.

Schlossman, S., Mershon, S., Livers, A., Jacobson, T., & Haggerty, T. (1993). Integrating Blacks into the U.S. military. In National Defense Research Institute, *Sexual orientation and U.S. military personnel policy: Options and assessment* (pp. 158–190). Santa Monica, CA: RAND.

Scull, K. C. (1990). *COHESION: What we learned from COHORT.* USAWC Military Studies Program Paper, Carlisle Barracks, PA: U.S. Army War College.

Shatzkin, K. (1993, May 30). Gay officer quits, cites harassment. *Seattle Times,* p. A1.

Shaw, M. E. (1976). *Group dynamics: The psychology of small group behavior.* New York: McGraw-Hill.

Sherif, M., Harvey, O. J., White, B. J., Hood, W. R., and Sherif, C. W. (1961). *Intergroup conflict and cooperation: The Robber's Cave experiment.* Norman, OK: University of Oklahoma Press.

Siebold, G. L., & Kelly, D. R. (1988). *Development of the combat platoon cohesion questionnaire.* Alexandria, VA: U.S. Army Research Institute for the Behavioral and Social Sciences.

Siebold, G. L., & Lindsay, T. J. (1994, August). *The relation between soldier racial/ethnic group and perceived cohesion and motivation.* Paper presented at the annual meeting of the American Sociological Association, Los Angeles.

Stein, A. A. (1976). Conflict and cohesion: A review of the literature. *Journal of Conflict Resolution, 20,* 143–172.

Steiner, I. D. (1972). *Group process and productivity.* New York: Academic Press.

Stephan, W. G. (1985). Intergroup relations. In G. Lindzey & E. Aronson (Eds.), *The handbook of social psychology* (3d ed., Vol. 2, pp. 599–658). New York: Random House.

Tajfel, H., & Turner, J. C. (1979). An integrative theory of intergroup conflict. In W. G. Austin & S. Worchel (Eds.), *The social psychology of intergroup relations* (pp. 33–47). Monterey, CA: Brooks/Cole.

Triandis, H. C. (1977). *Interpersonal behavior.* Monterey, CA: Brooks/Cole.

Tziner, A., & Vardi, Y. (1982). Effects of command style and group cohesiveness on the performance effectiveness of self-selected tank crews. *Journal of Applied Psychology, 67,* 769–775.

Wilder, D. A. (1986). Social categorization: Implications for creation and reduction of intergroup bias. In L. Berkowitz (Ed.), *Advances in experimental social psychology* (Vol. 19, pp. 291–355). New York: Academic Press.

Zaccaro, S. J., & McCoy, M. C. (1988). The effects of task and interpersonal cohesiveness on performance of a disjunctive group task. *Journal of Applied Social Psychology, 18,* 837–851.

The Deconstruction of Stereotypes:
Homosexuals and Military Policy

Theodore R. Sarbin

In 1986, in the wake of the prosecution of the Walker spy ring, the Department of Defense assigned the Personnel Security Research Center the task of determining whether biographical data, or biodata, of personnel cleared for access to classified information could be predictive of security risk. The biodata of interest included financial habits, drug and alcohol abuse, mental/emotional status, and sexual orientation. Captain Kenneth Karols and I were assigned to look into possible connections between sexual orientation and security risk. The task was not motivated by any interest to change the policy of excluding homosexuals from military service. Because security issues often overlapped with suitability issues, we enlarged the scope of our study to determine whether a nexus existed between homosexual orientation and suitability to serve in the armed forces.

After reviewing the available literature on homosexuality, we submitted to the Department of Defense a draft report. We recommended that research be undertaken the results of which might facilitate the integration of gay men and lesbians into the armed forces. The recommendation

This chapter is based on a paper delivered at the symposium *Integrating lesbians and gay men into the military*, 101st annual convention of the American Psychological Association, August 23, 1993, Toronto, Canada. I am grateful to Ralph Carney, Peter Lenrow, Mark Eitelberg, and Martin Wiskoff for helpful suggestions. The opinions expressed in this paper are those of the author and should not be construed as representing the position or policy of any department or agency of the U.S. government.

followed from a social psychological and historical analysis, and from the expectation that continuing pressures from civil rights advocates would influence the president, Congress, or the courts to rescind the discriminatory policy. On the basis of our conclusions that stereotypes, rather than empirically confirmed facts, provided the support for the resistance to changing the exclusionary policy, the recommended research would have focused on ways and means of deconstructing such stereotypes (Sarbin and Karols, 1989). Our draft report was rejected on the grounds that its scope went beyond the assignment of determining only whether a nexus existed between being homosexual and being a security risk.[1]

In the following pages, I discuss the nature and process of stereotyping and the conditions for modifying or eliminating stereotypes of gay men and lesbians. *Stereotype* as used in this paper carries the connotation of negative judgments of a class of people, often in the absence of any direct experience with members of the class. I am aware that the literature on the subject shows that stereotypes can be made up of positive attributions as well. Elsewhere I have made the claim that *social type* would serve as a better descriptor for the beliefs and attitudes we build up about classes of people, stereotype being a subset of social types (Sarbin, 1968). For the purposes of this chapter, I am following the current common usage: *Stereotype* refers to a set of beliefs and attitudes, predominantly negative, a subset of the general category of social type.

Because the conclusions in our study centered on the claim that negative stereotypes served as the basis for the exclusionary policy, I provide a brief historical account of the origin of the stereotypes of gay men and lesbians, leading to a discussion of the issues surrounding the policy of the military establishment. To bring the issues into the contemporary scene, I discuss a recent study that throws some light on the 1993 "Don't Ask, Don't Tell" policy.

The Social Construction of Stereotypes

The formation of stereotypes can best be understood in the language of social construction (Sarbin and Kitsuse, 1994). A review of social history makes clear that homosexuality has been rhetorically constituted in different ways at different times. As in the case of any social construction, the construction of homosexuality was created, negotiated, and modified to meet political and social goals. Four discrete constructions can be identified: homosexuality as sin, as crime, as sickness, and, most recently, as the defining feature of a minority group. Although each construction has its origins in different historical periods, derivatives of the earlier constructions continue to serve as a basis for the stereotypes that have guided military policy in the past and continue to guide policy under the plan adopted in 1993. Briefly, I review the four constructions.[2]

The origin of the sin construction goes back at least to biblical times (Boswell, 1980; Bullough, 1976; Greenberg, 1988). In the context of the dichotomy between good and evil, the conception of sin as transgression of divine law served as a powerful deterrent to nonconforming conduct. Historians point to two factors in the sin construction as applied to same-sex behavior. The first reflects the high priority placed on producing children. Because of high infant mortality, nonprocreational sexual behavior endangered the intergenerational continuity of the social group. Judeo-Christian religious precepts condemned homosexual behavior, masturbation, and any other sexual conduct that did not serve procreation (Bullough, 1976; Law, 1988). The second factor reflected the inferior status assigned to women. The attribution of inferiority was extended to men who took the female role in sexual relations. Historical accounts are relatively silent in regard to placing female homosexuality on the good-evil dimension.

Fundamentalist religious groups continue to use the vocabulary of sin to denote nonconforming sexual conduct. Biblical passages are quoted to support the negative stereotype, such as, "And if a man lie with mankind, as with womankind, both of them have committed an abomination: They shall surely be put to death; their blood shall be upon them" (Leviticus, 20:13). In political circles, contemporary advocates of this construction tend to avoid using the affect-laden religious term. Instead, they employ the somewhat less ecclesiastical language of morality and refer to homosexual relations as immoral.[3]

The second construction, homosexuality as crime, was an extension of the sin concept. About the sixteenth century, when the church surrendered to secular authorities the task of regulating conduct, some actions that had been classified as sin were redefined as crimes, punishable by the state. The frequently used appellation *sins against nature* was transformed to *crimes against nature* (Hoffman, 1968). A semiotic analysis of the use of the word "nature" in these constructions leads to the inference that the term is intrinsically ambiguous. Unlike the use of "nature" to refer to mountains and lakes, earthquakes and floods, birds and bees, the expression "crimes against nature" is an oblique way of referring to acts that violate social norms.

The third construction, homosexuality as sickness, developed within the last century as a concomitant of the medicalization of deviance. The success of science and technology in the nineteenth century influenced physicians to construct elaborate theories to account for all kinds of nonconforming behavior. Conduct that had been the province of moral and legal entrepreneurs was reinterpreted as illness. Textbooks of psychiatry and psychology asserted that nonconforming sexual conduct, in the absence of causal conditions such as microbes, brain disease, or trauma, was a manifestation of *mental* illness (see, e.g., Noyes and Kolb, 1958). Because a diagnosis of homosexuality could not meet standard criteria for sickness or disease (Gonsiorek, 1991), this construction no longer has official sanction: In the 1970s the American Psychological Association, the American Sociological Association, and the American Psychiatric Association issued unequivocal statements that homosexuality was not a mental disorder (Bayer, 1981).

The fourth construction, the minority group construction of homosexuality, follows from considering the definition of a minority group: persons who share the experience of being the objects of discrimination, exclusion, and persecution by members of nonminority groups who hold ethnocentric beliefs and stereotypes (Paul, 1982; Sagarin, 1971). The construction that identifies sexual orientation as a feature of a minority group reflects the historical period in which tolerance of differences took on political colorings.[4] The civil rights movement improved the status of ethnic and racial minorities before the law. Moreover, the civil rights movement had increased society's tolerance of diversity, including, for example, the removal of gender barriers to employment. Another outcome was the sometimes reluctant recognition that gay men and lesbians could perform an infinite variety of occupational and recreational roles in the civilian sector. Opinion polls indicate a slow but definite increase in tolerance for homosexual persons. For example, recent public opinion polls show that half the respondents agree that homosexuals should be allowed to serve in the armed forces (see, e.g., Moskos, 1993). The minority group construction now competes with the earlier constructions for public acceptance. In some jurisdictions, the minority group construction has been further legitimized by the enactment of legislation that guarantees equal access to housing and jobs.

The content of, and the reasons for, stereotypes are a mixture, an amal-gam of beliefs and values. From the arguments advanced by politicians and military officers who support the ban of gays and lesbians from the military, it is clear that they hold stereotypes that are derived from the sin construction. In the 1993 congressional hearings and in public com-mentaries, hardly anyone cited the sickness construction, and only those with a legalistic turn of mind invoked the crime construction. Although consensual homosexual relations between adults is no longer a crime in most Western countries, it remains on the statute books of half the states in this country and in the U.S. Uniform Code of Military Justice. When supporters of the exclusionary policy invoke the "crime against nature" construction, the context makes clear that it is a rhetorical device for reinforcing the attribution of sin or immorality to homosexual men and women.

From the recent and ongoing debates, then, it is reasonable to infer that the major friction is between the sin construction and the minority group construction. No argument is made that gay men and lesbians could not function competently as war fighters or peacekeepers. Rather, the argument is advanced that heterosexual personnel would be uncom-fortable or would feel threatened under conditions of barracks and ship-board life if acknowledged homosexual men and women were allowed to serve. Feeling uncomfortable or threatened, the argument continues, would lead to a breakdown in "cohesion," a necessary ingredient for war fighters (see chapter 8). The obsessive concern with sharing showers and sleeping quarters is vocalized in terms of privacy and modesty values (see chapter 11). The same arguments were proposed when President Truman ordered the integration of African Americans into the military (see chap-ter 5).

What is the source of this expression of the perception of threat? The sin construction places a stigma on the gay man or lesbian. The stigma is not visible as in the case of sewing a badge of dishonor on the sinner. My interviews (Sarbin, 1991) with military personnel support the hy-pothesis that the suspected or acknowledged homosexual person is per-ceived as a polluted specimen, the pollution forming an invisible and si-lent miasma (Douglas, 1966). In the course of my interviews with Navy officers, nearly all said that they would be willing to share work assign-ments with a fellow officer who acknowledged being homosexual. How-ever, they added that they would not be comfortable sharing sleeping quarters with a gay man. Asked if they were fearful about being sexually assaulted, they replied that they had no such fears. Queried further, the officers would give no explicit reason save for feeling uncomfortable. Further discussion led to the inference that somehow the space would be polluted, not by germs, but by an unarticulated conception of the gay man as a tabooed object and a carrier of sin.

Motivation for Stereotyping

Stereotypes are constructed from exaggerations of social types that feature some negatively valued conduct such as style of speech, accent, demeanor, or expressive gestures. The core of the common stereotype of gay men is that of the feminized male; for lesbians, it is the masculinized female (Bell, 1973; Williams and Weinberg, 1971; Bell and Weinberg, 1978). The stereotype, in the first instance, gets its energy from the crossing of gender categories. For persons who have been enculturated to the strict functional separation of gender, an individual who crosses gender lines is guilty of breaching a fundamental moral rule (Karst, 1991). The stereotype is activated, for example, when a man adopts gestures that are traditionally regarded as feminine.

Once a stereotype is formed and the empirical fact of individual differences is ignored, the negative characteristics assigned to the target class become elaborated (Hamilton, 1979; Bar-Tal, 1989). A feature of the elaborated version is that homosexuals are one-dimensional—that their conduct is predominantly directed by sexual desire. Aided and abetted by selective media images, the stereotype persists in the face of a great deal of evidence that the expressive behavior of most homosexual persons does not match the image of the feminized male or the masculinized female. Further, the evidence of numerous studies makes clear that the degree of sexual interest among gay men and lesbians does not match the one-dimensional feature of the stereotype (Bell, 1973). In the 1993 congressional hearings, military experts advanced opinions that implied the underlying and unspoken unidimensionality of the stereotype—that homosexual men and women, as a class, are sexual predators. As in other instances of negative stereotyping, the exclusionary and even hostile response to gay men and lesbians follows, not from an assessment of behavior, but from assigning them to suspected or actual membership in a broadly defined out-group—homosexuals.

To understand and to deconstruct stereotypes, it is first necessary to grasp why people hold them in the first place. We need to know not only the content of stereotypes, but also the *reasons* that some politicians, some military officials, and some rank and file personnel hold the stereotypes that guide their exclusionary behavior. The question is not what *causes* people to construct and use stereotypes as much as the *reasons* for holding and acting on stereotypes. A recent analysis by Snyder and Miene (1993) is helpful. Consistent with a constructionist perspective, their formulation attempts to show the reasons that support the construction and use of stereotypes. Important to their argument is the premise that the person holding the stereotype is goal-directed, an agent capable of making choices. They identified three classes of reasons. The first, to achieve cognitive economy, is a general motive for any social typing. In everyday life, we construct and use typologies to classify people as cat lovers or bird lovers, bookworms or jocks, serious or fun-loving, and so forth, without necessarily assigning valuational properties (Sarbin, 1968). The second class has to do with ego enhancement, what people do and say to create or bolster an acceptable identity. The third class comprises those reasons that facilitate a person's becoming a fully socialized member of a particular group. Becoming a member of a defined cultural group involves not only the acquisition of speech and gesture patterns, rules of etiquette, style of dress, and so forth, but also implicit rules for determining who is an insider and who is an outsider. In the process, he or she creates social objects, some of whom are regarded as benign and some of whom are excluded and rejected.

I have been conducting informal interviews with men who had been socialized during adolescence to reject gays and lesbians, but who, as mature adults, had taken a nonprejudicial posture (Sarbin, 1994). Although some respondents indicated a kind of social comparison process in the interest of ego enhancement, it is Snyder and Miene's third class of reasons that appears to have had the major effect in the initial development of the stereotype. That is, the stereotype appears to have been constructed through various forms of socialization. Primarily during the adolescent years, my informants absorbed beliefs about what social objects to accept and what social objects to reject, even to hate. As is the case with many beliefs and practices, the reasons for acquiring discriminatory and exclusionary practices centered on peer acceptance. As young adolescents, even before they had well-formulated ideas about sexuality in general and same-sex orientation in particular, they had acquired rhetorical techniques for assigning nonperson status to homosexuals through hostile jokes, salacious stories, and pejorative labeling.

The adult effects of adolescent enculturation were illustrated in interviews conducted by television reporters during the 1993 congressional hearings on homosexuals in the military. Some soldiers and sailors supported the discriminatory policy. Asked to explain why they would agree with continuing the ban against lesbians and gay men, the sin construction was invoked with remarks such as "What they do is immoral." For those whose enculturation had included the authority of religious scriptures, the Bible was cited as further justification for intolerance and/or hatred.

What is acquired is the practice of making negative judgments about people solely on class membership, such as, "All Scots are frugal," "All women are emotional," or "All Russians are humorless." Such judgments require a minimum of cognitive work. The person making the judgment is not required to make a personality assessment, to observe the actual conduct of the target individuals, or, in fact, to have any direct commerce with members of the negatively valued class.

The significant feature in stereotyping is the form of the silent major premise. In Aristotelian rules of logic the formulation would be: All x are y. If a person is a member of the class x, then that person is simultaneously a y. In the controversial accession policy of the military, the major premise is "All homosexuals are troublesome." Therefore, any individual identified as homosexual is automatically categorized as troublesome. The recognition of the uncritical use of the quantitative term "all" in the major premise could influence the formulation of proposals to deconstruct stereotypes that support the military's policy. Persuasive communications can be formulated that would shift the form of the major premise from "*All* homosexuals are troublesome" to "*Some* homosexuals are troublesome" (with the implication that "*Some* homosexuals are not troublesome.") With this qualified premise, no conclusion is possible for any particular person. He or she could be a member of the subsample who would contribute to poor morale, or a member of the subsample who would have a benign influence on morale. Using the "Some x are y" formulation, then, demands cognitive work on the part of the person making the judgment. He or she would have to go beyond the simplistic method of excluding (or including) a person by virtue of his or her minority group status. Rather, the person making the judgment would have to ask and answer other questions to determine whether the target person, not the class, should be identified as acceptable. As is well known, many gay men and lesbians have served honorably and effectively in the military services. Using the "Some *x* are *y*" formula would render sexual orientation irrelevant to becoming a functioning member of the armed forces.

In formulating conclusions about my informants who no longer hold the stereotype, it appears that they had undergone a resocialization process. In some cases, the stereotype dissolved as part of a general increase in the tolerance for diversity as a result of the diffusion of the persuasive communications of the civil rights movement. In other cases, the stereotype dissolved following the recognition that a relative or a respected coworker was homosexual. The informants replaced the "*All x* are *y*" formula with a formula more consistent with practical reasoning, that is, "*Some x* are *y*." A review of my informants' narratives suggested that they had formed nonexclusionary beliefs about homosexuals through a socialization process that proceeded from value orientations more complex than adolescent peer acceptance. Whether to accept or reject a person would depend upon dimensions other than sexual orientation, such as honesty, competence, stamina, courage, trustworthiness, helpfulness, and cooperation.

Deconstructing Stereotypes

Having identified some of the sources, how would one go about the task of deconstructing firmly held stereotypes among military personnel? The following sketch of a proposal draws upon two traditions in social psychology: the theory of social influence advanced by Kelman (1974) and the corpus of research on prejudice associated with Allport (1954), Pettigrew (1986), Sherif, Sherif, and Nebergall (1965), and Cook (1978), among others. Specifically, I make use of the concept of *compliance* from Kelman's theory and the *contact hypothesis* as advanced by the researchers on prejudice.

Kelman proposed three processes of social influence: compliance, identification and internalization. In compliance, persons publicly perform acts that may be contrary to their private beliefs because they could be punished for noncompliance or rewarded for compliance. In identification, persons take on the attitudes of others whom they regard as mentors or role models. In internalization, public acts conform to privately held beliefs and values.

The distillation of the research on prejudice allows the inference that "contact" with the objects of prejudice reduces prejudicial attitudes, especially if the contact is in the context of shared work assignments (see chapter 10). One explanation for this phenomenon is that contact makes it possible for the persons holding the stereotype to make judgments about the objects of prejudice as individuals, rather than as members of a class. In so doing, they use the same dimensions of judgment that they would use to evaluate the personalities or conduct of nonstigmatized persons, dimensions unrelated to sexual orientation.

The organization and traditions of the military make possible the utilization of Kelman's compliance model. The basis of military discipline is giving and following orders. Military discipline was a potent factor in the achievement of racial integration in the services. Integration was not achieved overnight; many white soldiers and sailors, including officers, were reluctant to cross the color barrier in eating and housing arrangements. Nevertheless, they ultimately complied with the order. In this connection, in informal face-to-face interviews, I have asked the following question of enlisted personnel who favor retaining the ban against homosexuals, "What would you do if your commanding officer issued an order for you to participate in a dangerous mission with a comrade whom you knew to be gay?" The almost uniform response was, "I would follow orders" (Sarbin, 1992). A similar item was included in a recent attitude survey of Navy officers (Cleveland and Ohl, 1994). Fifty percent of the respondents indicated they would have "no difficulty obeying an order from a commanding officer to work with a homosexual coworker on a difficult or dangerous assignment." (It is interesting to speculate about the proportion of officers who would have answered the same way, say, twenty years ago. I would guess that the proportion would have been much smaller, so 50 percent would indicate a softening of negative attitudes.)

Compliance, that is, following orders, was one of the conditions that led to dissolving racial stereotypes in the military. The orders came from the commander in chief. The present status of African Americans in the military is evidence of the changes occurring over the years. Central to compliance is the expectation of nondiscriminatory overt conduct. Kelman's other forms of social influence—identification and internalization of egalitarian attitudes—in many instances appear to have followed habitual compliant actions.

The organization of the military provides an opportunity for changing attitudes and beliefs through compliance. Models exist: current training and public relations procedures to eliminate sexual harassment, and, as I have suggested, earlier procedures to deal with racial prejudice.

Influence of Construing the Military as a "Job"

Department of Defense studies make clear that in the all-volunteer force, the major motives for enlistment are employment and opportunities for vocational training (Bowman, Little, and Sicilia, 1986). For most contemporary personnel, entering the military is similar to entering a job. The metaphor of "job" comes up repeatedly in the accounts of service members. Enlisted personnel during the Persian Gulf War responded to the queries of television reporters with the same language: "We have a job to do here."

The job metaphor is pertinent to the model I have been developing. The criteria employed in evaluating persons in a job setting involve competence. On this reasoning, once a person is perceived as competent in performing his or her duties, stereotypic criteria such as gender, race, or sexual orientation fade into the background and ultimately become irrelevant.

A news item illustrates the stereotype-abatement effects of working together. In this case, the issue centered not on gay men and lesbian women, but on another minority group: women in the military. A reporter queried nine high-ranking noncommissioned officers who had nearly two hundred years of naval service among them, mostly as submariners. To the question "Should women be allowed to serve on submarines?", the answer was a unanimous "Yes." Supporting the affirmative answer, a master chief for a submarine squadron told a story about women who maintain the submarines from the Navy's fleet of gender-integrated submarine tenders. "We once had some main engine bearings that had to be taken off the ship. . . . The men on my ship couldn't do it, and guess who did: two women. They rigged them right out of the ship in an hour and the men stood there with their mouths open. . . . You see it every day. That's why there's no contention that the capabilities to do the job are the same between men and women" (Pexton, 1993, p. 18).

In sum, being able to demonstrate competence influences acceptance. Having the opportunity to demonstrate competence would be a first step toward dissolving the stereotype. To make judgments about persons performing their duties requires criteria that are unrelated to gender or to sexual orientation. That gay men and lesbians have demonstrated competence is attested by the military records of dismissed personnel who appealed to the courts and whose biographies appeared in the media: Keith Meinhold, Margarethe Cammermeyer, Tracy Thorne, Dusty Pruitt, and Joseph Steffan, among others. That acknowledged homosexual personnel can function effectively without damaging morale is dramatically demonstrated in those cases in which the courts have ordered the military to reinstate petitioners who were dismissed solely for announcing their sexual orientation. Besides demonstrating competence on the job, gay and lesbian personnel have demonstrated gallantry, bravery, and valor, characteristics highly valued in the military community (Bérubé, 1990; Shilts, 1993; Humphries, 1990).

The New Policy: "Don't Ask, Don't Tell"

The social context in which competence is demonstrated must be considered, including clearly articulated communications from military commanders. In the same way that commanders—through the chain of command—are being made responsible for, and instrumental in, resocializing their troops about sexual harassment issues, commanders could be educated about their responsibility for resocializing their troops in the invalidity of previously reinforced gay and lesbian stereotypes.

The issue of responsibility of commanders is especially pertinent, given the recent promulgation of the policy that was announced as a compromise between legislators who favored keeping the ban, and President Clinton, who favored lifting the ban. The "Don't Ask, Don't Tell" policy (Qualifications Standards for Enlistment, Appointment and Induction, 1993) removed the gatekeeping function from recruitment officers. Before implementation of the directive, commanders could assume that all homosexuals had been screened out. If a man or woman were identified as a homosexual, it was seen as a failure of the recruitment and screening process and steps could be taken immediately to separate the gay man or lesbian from military service. The present policy places the burden of enforcement on the commander, and as previously stated, the attitudes and practices of commanders would be important variables both in applying the new policy and, perhaps in the future, participating in practices that would dissolve the negative stereotypes.

A recently completed study throws some light on how the new policy is perceived by active-duty military officers (Cleveland and Ohl, 1994).[5]

The attitudes of naval officers who were enrolled in graduate programs at the Naval Postgraduate School in Monterey, California, were assessed in the spring of 1994. Of the 1,000 self-administered questionnaires distributed to the entire student body, 605 were completed and returned to the investigators. The respondents were instructed not to sign their names. Fifty officers wrote qualitative comments about the issues covered in the questionnaire. In addition, twenty officers were interviewed face-to-face to gather additional in-depth information about the origins of their attitudes.

The intent of the study was to determine whether the officers understood the directive and the extent to which their attitudes and beliefs might complicate their new roles as gatekeepers. The self-administered questionnaires and the face-to-face interviews also provided information on a number of related issues, such as differences in attitudes associated with gender, years in service, and being acquainted with a homosexual person.

Central to the directive is the distinction between conduct and orientation. According to the directive, one's sexual orientation is a private matter. One's conduct, however, is subject to review and may be the basis for separation. Problems emerge because of the directive's unique definitions of conduct and orientation. In ordinary language, conduct is equated with overt behavior, with performance, with doing, whereas orientation is an abstract entity, the identification of a status. The directive furnishes examples of both conduct and orientation. Military commanders are informed, for example, that socializing in a gay bar is not to be considered conduct and is not a basis for instituting an investigation. Everyday users of the English language, however, would consider such an act to be conduct. On the other hand, everyday users of the English language would not regard as homosexual conduct a person telling an officer of his or her sexual orientation; under the directive, such an admission is considered conduct, and the officer would be required to institute proceedings for separating the person from the service. The tortured logic of the compromise built into the directive is reflected in the responses to questions designed to assess how well the officers know the new rules.

Following are items from the self-administered questionnaire, the responses to which reflect confusion in interpreting the new policy:

"If a service member tells a superior that he/she has a homosexual orientation, this is equivalent to sexual misconduct." The options "dis-

agree" or "strongly disagree" were checked by 73 percent of the respondents. That is, nearly three-fourths of the respondents continued to follow ordinary English usage, contrary to the instructions in the directive.

"*Lawful off-duty sexual activity would be of no concern to me.*" The options "agree" or "strongly agree" were checked by 70 percent of the respondents. Thus, a large proportion of the officers expressed attitudes that are contrary to the instructions in the directive that state that off-duty sexual conduct is grounds for investigation and separation.

"*As a department head, you receive a report from Seaman Smith that Airman Jones was holding hands with a same-sex civilian in a movie theater. It is your responsibility to investigate this activity.*" The options "disagree" or "strongly disagree" were checked by 56 percent of the respondents. That is, more than half the officers held beliefs that are contrary to the directive that hand-holding is considered equivalent to conduct.

The sample studied is probably representative of the officer corps. The military culture tends to foster conservative attitudes, as reflected in the responses to the following item:

"*I would prefer not to have homosexuals in my command.*" The options "agree" or "strongly agree" were checked by 82 percent of the respondents.[6]

The prevalence of conservative attitudes is not unexpected, given the culture of the military. Despite the conservative views of the respondents, as many as 30 percent "would have no difficulty working with a homosexual commanding officer." Thirty-eight percent agreed with the statement, "A division officer's sexual preference has no effect on the officer's ability to lead." Almost 50 percent agreed with the statement, "I would have no difficulty obeying an order from a commanding officer to work with a homosexual coworker on a difficult/dangerous assignment." And 60 percent agreed with the statement, "Homosexuals and heterosexuals should have equal rights."

One would have predicted more uniform results, given the expectation that military officers by and large subscribe to conservative values. An ambivalence is suggested by the observation that from 30 percent to 60 percent of the officers endorsed items that are inconsistent with the conservative viewpoint and with the implications of the new policy. One can interpret these findings as a softening of officers' rigidly held views on homosexuality, consistent with changes in civil society. In support of this interpretation, half the respondents agreed with the statement, "It is just a matter of time until military policy is changed to full and open acceptance of homosexuals."

This inference is based on data from both the self-administered questionnaire and the interviews: Officers' attitudes are characterized by ambivalence, a conflict between conservative attitudes and the desire to be fair. An excerpt from one of the interviews is typical:

> I'm somewhat ambivalent [about the new policy]. I personally don't respect homosexuals. I think that type of activity is wrong . . . not so much from a religious standpoint, although I can understand why . . . from a moral standpoint, it degrades the moral fiber of our society. Until those conditions change, that's how I'll think about it. On the other hand, we live in America . . . America is certainly founded on the premise that people's rights are primary . . . people have a right to live their life the way they want to. I struggle with that quite a bit—the way I feel personally and people's rights. Once you start taking away people's rights you end up with a communist philosophy. (Cleveland and Ohl, 1994, p. 93)

One of the possible outcomes of the ambivalence is confusion and inconsistency in implementing the "Don't Ask, Don't Tell" policy.

Responses to the survey and data from the interviews lend support to the contact hypothesis described earlier as one of the avenues for dissolving stereotypes. To the item, "I have a friend or relative who is homosexual," 29 percent answered "yes," 19 percent answered "possibly," and 52 percent answered "no." These results were cross-tabulated with another item, "I feel uncomfortable in the presence of homosexuals and have difficulty interacting normally." Of those respondents who indicated they had a homosexual friend or relative, 67 percent indicated they would *not* feel uncomfortable. Of those who answered "possibly" to having a homosexual friend or relative, 54 percent would *not* feel uncomfortable. These results are consistent with the contact hypothesis. These numbers represent a modest change in the military culture, given its traditional conservatism.

About 10 percent of the respondents were women officers, consistent with the proportion of women in the Navy as a whole. Several items in the survey results indicate a more tolerant attitude toward homosexuals; for example, 40 percent of the women but only 14 percent of the men would have no objection to having homosexuals in their commands. Three-fourths of the women, but fewer than half the men, would be willing to work with a homosexual coworker on a dangerous assignment. More than three-fourths of the men, but about half the women, agree that allowing homosexuals into the military services sends the wrong message to society. Looking ahead at the gatekeeping function now assigned to junior officers, the disparity between male and female officers may contribute to inconsistent implementation of the directive.

The data were analyzed according to length of time in the Navy, which is highly correlated with age, given the "promote or perish" policy. It is readily apparent that younger officers (those who have been in the Navy two to nine years) indicated more tolerant attitudes than older officers (those who have been in the Navy ten to fifteen or more years). The data offer no clues in regard to the reason for this finding. Have the older officers had more time to acquire and reinforce the conservative values of the military? Or are the younger officers members of an age cohort that holds more tolerant attitudes?

It is clear from this study that Naval officers have not yet assimilated the special language of the "Don't Ask, Don't Tell" policy. This fact, together with the variable degrees of tolerance among the officers, suggests that the implementation of the policy is likely to be characterized by inconsistency and conflict.

Data gathered in 1995 indicate that the policy has apparently had no effect on the proportion of gay men and women who are dismissed from the military services (Moss, 1995; Pine, 1995; Schmitt, 1995). Data gathered in 1996 make clear that discharges for homosexuality have increased significantly.* That the proportion of such discharges has not decreased, contrary to expectations, may be due to difficulties inherent in interpreting the ambiguous language of the directive, one outcome of which is the continued use, by default, of traditional negative stereotypes.

Summary

Homosexuality is a social construction. Of the four constructions still active, insofar as the military is concerned, the conflict is between the sin construction and the minority group construction. The sin construction has led to the formation of negative stereotypes of gay men and lesbian women.

In principle, the military could assimilate homosexuals by employing

---

*Data supplied by Defense Manpower Data Center.

the results of social psychological research to dissolve negative stereotypes. The "Don't Ask, Don't Tell" policy does not require programs to change attitudes and beliefs of members of the military. But the findings in the survey suggest that attitudes are changing in favor of more tolerance and acceptance. This shift is probably a reflection of the more tolerant attitudes in the wider society of which the officers and their families are members.

The results of the survey of Naval officers make clear that knowledge of the new policy is variable, that attitudes are likewise variable, and that ambivalence characterizes the issue. Such ambivalence and variation in knowledge and attitudes are likely to result in inconsistent implementation of the policy and the persistence of practices associated with negative stereotypes.

Notes

1. During the period when the draft report was under review by Department of Defense officials, an anonymous person leaked a copy to members of Congress, who ultimately passed it on to the press. Subsequently, since the report had entered the public domain and had not been classified, a commercial publisher undertook to publish it (Dyer, 1990).

2. Ralph Carney pointed out that the first three constructions were formed by persons other than homosexuals. The minority group construction was initiated by lesbians and gay men.

3. Not all politicians avoid "sin" in phrasing their arguments. In discussions about expert testimony given to the House Armed Services Committee in May 1993, Representative Robert Dornan (R-CA) said: "all of the psychologists and psychiatrists in the world are never going to convince me that heterosexual sex . . . [outside marriage] . . . is [not] a sin, and I am not going to cut a different standard" for homosexual sex (Policy implications of lifting the ban, 1993, p. 303).

4. In the late 1960s and early 1970s, a short-lived movement influenced the development of the minority group construction. The dominant paradigm was a "liberation" concept: If everyone would openly express his or her heterosexual and homosexual feelings, the categories "heterosexual" and "homosexual" would fade away. By the end of the 1970s, the minority group construction became dominant in the gay community.

5. The study was conducted by Commander Fred Cleveland and Commander Mark Ohl to fulfill the requirements for the master's degree at the Naval Postgraduate School. The research was supervised by Professor Mark Eitelberg and the author.

6. An alternate interpretation is possible. The responses might indicate a pragmatic desire to avoid the difficulties associated with being gatekeepers and applying fairness criteria in implementing the exclusionary policy.

References

Allport, G. (1954). *The nature of prejudice.* Cambridge, MA: Addison-Wesley.

Bar-Tal, D. (Ed.). (1989). *Stereotypes and prejudice.* New York: Springer-Verlag.

Bayer, R. (1981). *Homosexuality and American psychiatry: The politics of diagnosis.* New York: Basic Books.

Bell, A. P. (1973). Homosexualities: Their range and character. *Nebraska Symposium on Motivation, 21,* 1–26.

Bell, A. P., & Weinberg, M. S. (1978). *Homosexualities: A study of diversity among men and women.* New York: Simon & Schuster.

Bérubé, A. (1990). *Coming out under fire: The history of gay men and women in World War II.* New York: Free Press.

Boswell, J. (1980). *Christianity, social tolerance, and homosexuality.* Chicago: University of Chicago Press.

Bowman, W., Little, R., & Sicilia, G. T. (Eds.). (1986). *The all volunteer force after a decade.* Washington: Pergamon Brassey's.

Bullough, V. (1976). *Sexual variance in society and history.* Chicago: University of Chicago Press.

Cleveland, F. E., & Ohl, M. A. (1994). *"Don't ask, don't tell"—policy analysis and interpretation.* Unpublished master's thesis, Naval Postgraduate School, Monterey, CA.

Cook, S. (1978). Interpersonal and attitude outcomes in cooperating interracial groups. *Journal of Research in Development and Education, 12,* 97–113.

Department of Defense. (1993). *Qualifications standards for enlistment, appointment and induction.* (Directive 1304.26). Washington, DC: U.S. Government Printing Office.

Douglas, M. (1966). *Purity and danger: An analysis of the concepts of pollution and taboo.* London: Routledge & Kegan Paul.

Dyer, K. (Ed.). (1990). *Gays in uniform.* Boston: Alyson Press.

Gonsiorek, J. C. (1991). The empirical basis for the demise of the illness model of homosexuality. In J. C. Gonsiorek & J. D. Weinrich (Eds.), *Homosexuality: Research implications for public policy* (pp. 115–136). Newbury Park, CA: Sage.

Greenberg, D. F. (1988). *The construction of homosexuality.* Chicago: University of Chicago Press.

Hamilton, D. L. (1979). A cognitive attributional analysis of stereotyping. In L. Berkowitz (Ed.), *Advances in experimental social psychology* (Vol. 12, pp. 53–84). New York: Academic Press.

Hoffman, M. (1968). *The gay world.* New York: Basic Books.

Humphries, M. A. (1990). *My country, my right to serve.* New York: HarperCollins.

Karst, K. L. (1991). The pursuit of manhood and the desegregation of the armed forces. *UCLA Law Review, 38,* 499–581.

Kelman, H. (1974). Further thoughts on the processes of compliance, identification, and internalization. In J. T. Tedeschi (Ed.), *Perspectives on social power* (pp. 125–171). Chicago: Aldine.

Law, S. A. (1988). Homosexuality and the social meaning of gender. *Wisconsin Law Review, 88,* 187–235.

Moskos, C. (1993, January 31). Soldiering: It's a job, not an adventure in social change. *Washington Post,* p. C3.

Moss, J. K. J. (1995, March 24). Military may not be meeting intent of "don't ask, don't tell." *Washington Times,* p. 3.

Noyes, A. P., & Kolb, L. C. (1958). *Modern clinical psychiatry* (5th ed.). Philadelphia: W. B. Saunders.

Paige, K. E. (1977). Sexual pollution: Reproductive sex taboos in American society. *Journal of Social Issues, 33,* 144–165.

Paul, W. (1982). Minority status for gay people: Majority reaction and social context. In W. Paul, J. D. Weinrich, J. C. Gonsiorek, & M. E. Hotvedt (Eds.), *Homosexuality: Social, psychological and biological issues* (pp. 351–370). Beverly Hills, CA: Sage.

Pettigrew, T. F. (1986). The intergroup contact hypothesis reconsidered. In M. Hewstone & R. Brown (Eds.), *Contact and conflict in intergroup encounters* (pp. 169–195). Oxford: Blackwell.

Pexton, P. (1993, July 12). Women in subs? Men say yes. *Navy Times,* p. 18.

Pine, A. (1995, February 6). Few benefit from new military policy on gays. *Los Angeles Times,* p. 1.

Policy implications of lifting the ban on homosexuals in the military: Hearings before the Committee on Armed Services, House of Representatives, 103d Cong., 1st Sess. 1 (1993). Washington, DC: U.S. Government Printing Office.

Sagarin, E. (1971). *The other minorities: Nonethnic collectivities conceptualized as minority groups.* Waltham, MA: Ginn and Co.

Sarbin, T. R. (1968). On the distinction between social roles and social types, with special reference to the hippie. *American Journal of Psychiatry, 125,* 1024–1031.

———. (1991). [Interviews with naval officers]. Unpublished raw data.

———. (1992). [Interviews with military personnel]. Unpublished raw data.

———. (1994). *Dissolving stereotypes.* Unpublished manuscript, Defense Personnel Security Research Center, Monterey, CA.

Sarbin, T. R., & Karols, K. (1988). *Nonconforming sexual orientation and military suitability.* Monterey, CA: Defense Personnel Security Research Center.

Sarbin, T. R., & Kitsuse, J. I. (Eds.). (1994). *Constructing the social.* London: Sage.

Schmitt, E. (1995, March 13). The new rules on gay soldiers: A year later, no clear results. *New York Times,* p. 1.

Sherif, C. W., Sherif, M., & Nebergall, R. E. (1965). *Attitude and attitude change: The social judgement-involvement approach.* Philadelphia: W. B. Saunders.

Shilts, R. (1993). *Conduct unbecoming: Gays and lesbians in the U.S. military.* New York: St. Martin's Press.

Snyder, M., & Miene, P. (1993). On the functions of stereotypes and prejudice. In M. P. Zanna (Ed.), *The psychology of prejudice: The Ontario symposium* (Vol. 7, pp. 33–54). Hillsdale, NJ: Lawrence Erlbaum.

Williams, C. J., & Weinberg, M. S. (1971). *Homosexuals and the military.* New York: Harper & Row.

Why Tell If You're Not Asked? Self-Disclosure, Intergroup Contact, and Heterosexuals' Attitudes Toward Lesbians and Gay Men

Gregory M. Herek

The United States military's principal justification for its policies concerning homosexual personnel has very little to do with the actual abilities or characteristics of gay men and lesbians. The Department of Defense (DoD) has virtually abandoned its past arguments that homosexual men and women are psychologically impaired, a security risk, or incapable of performing their duties, and therefore are inherently unfit for military service (Herek, 1993). Instead, the DoD now concedes that lesbians and gay men can serve honorably and capably, and acknowledges that they have done so in the past. Indeed, the current policy ("Don't Ask, Don't Tell, Don't Pursue") allows service by gay people provided that they keep their sexual orientation a secret.

Thus, current policy is less about homosexuality than it is about heterosexuals' reactions to homosexuality and to persons who are gay or lesbian. The DoD argues that heterosexual personnel would be unwilling to work with or obey orders from a gay man or lesbian, that they would be unwilling to share sleeping quarters or latrines with them, and that the presence in a unit of an individual known to be gay would reduce cohesion and thereby impair performance (for elaboration on these points, see chapters 8 and 11). These arguments boil down to a concern about information: how gay people manage information about their sex-

I wish to thank Steve Franzoi, Rob MacCoun, Jack Dynis, and Jared Jobe for their helpful comments on earlier versions of this chapter.

ual orientation, and how heterosexuals react to information that another service member is gay.

In the present chapter, I review theory and research from the behavioral and social sciences to provide an understanding of the processes whereby gay people—as members of a stigmatized minority group—manage information about their status, how and why they disclose this information to others (popularly referred to as *coming out of the closet*, or simply *coming out*), and the effects of receiving such information on members of the heterosexual majority group. I also explain why self-disclosure about one's sexual orientation—undertaken either as an end in itself or incidentally to achieving other goals—is important for an individual's well-being, regardless of whether that individual is a heterosexual, homosexual, or bisexual. I argue that the current policy imposes unequal restrictions on the speech and conduct of different military personnel by placing minimal constraints on heterosexuals' disclosure of information about their sexual orientation while prohibiting gay men and lesbians from doing the same.

Because many issues to be addressed here concern social interaction and interpersonal disclosure, the chapter begins with a brief discussion of scientific research relevant to those topics. Next, basic aspects of sexual orientation are discussed as a prelude to considering self-disclosure about sexual orientation by heterosexuals. Then the asymmetries of experience between heterosexuals and homosexuals are described, followed by discussion of the reasons why gay men and lesbians—despite societal sanctions—come out to others. Then data are presented concerning the impact of such disclosure on heterosexuals. Finally, the implications of current military policy are considered in light of the foregoing discussion.

Self-Disclosure and Stigma

Coming out to another person is a form of *self-disclosure*, which is defined here as the communication by one individual to another of information about himself or herself that otherwise is not directly observable. By this definition, revealing one's height, weight, gender, or eye color does not usually constitute self-disclosure because such characteristics are apparent to the casual observer in most circumstances. In contrast, revelations about one's political beliefs, religious affiliation, personal income, family background, or sexual orientation would usually be classified as self-disclosure.

For the present discussion, four points about self-disclosure are particularly relevant. First, self-disclosure is an integral component of normal
social interaction. Even casual conversations with strangers typically involve self-disclosures about, for example, one's marital or parental status,
occupation, or opinions about a television program, sports team, or politician. An extensive body of research indicates that self-disclosure is an
integral component in the formation and maintenance of ongoing social
relationships with friends, coworkers, neighbors, and others (Altman and
Taylor, 1973; Derlega and Berg, 1987). Willingness to self-disclose is generally beneficial to one's social life and friendships, whereas patterns of
consistent nondisclosure are linked to loneliness and social isolation (e.g.,
Davis and Franzoi, 1986; Franzoi and Davis, 1985; Franzoi, Davis, and
Young, 1985; Stokes, 1987).

Second, self-disclosure can vary in its level of intimacy. Ongoing interpersonal relationships generally are closer to the extent that they involve
more intimate self-disclosures. Developing a relationship with someone—getting to know that person—is often analogous to the process of
peeling away the layers of an onion, with the uncovering of each layer
corresponding to revelations of progressively more intimate information
about the self. Such information is more personal or intimate to the extent that it (1) promotes broad generalizations about one's personality;
(2) distinguishes oneself from others; (3) reveals a characteristic that is
not readily observable to others; (4) reveals a characteristic that the larger
society regards as undesirable; (5) reveals a characteristic that may be
perceived as a vulnerability; and (6) is associated with high levels of emo-
tion or feeling (Archer, 1980; see also Altman and Taylor, 1973).

Third, the level of intimacy in a relationship is usually reciprocal, that
is, the parties in a relationship expect each other to share roughly equal
amounts of personal information, and to disclose information that is of
approximately the same level of intimacy (e.g., Berg and Derlega, 1987;
Derlega, Harris, and Chaiken, 1973). Lack of reciprocity in the intimacy
of self-disclosure—whether one party is perceived as disclosing too much
or not enough—is likely to strain a relationship (Fitzpatrick, 1987; Baxter, 1987).

Finally, the process of self-disclosure is complicated considerably for
people who possess a concealable stigma. As used here, *stigma* refers to
a pattern of serious social prejudice, discounting, discrediting, and discrimination that an individual experiences as a result of others' judgments
about her or his personal characteristics or group membership (e.g., Goff-

man, 1963; Jones et al., 1984). Whereas some stigmatized characteristics are readily visible to others (e.g., skin color, physical disability), others can often be concealed (e.g., membership in an ostracized religious or political group, homosexuality). Having a concealable stigma means that otherwise routine self-disclosures can place oneself at heightened risk for negative sanctions, that others are likely to regard such disclosures as highly—often inappropriately—intimate, and that reciprocity of disclosure is difficult to maintain in a personal relationship.

In the most influential theoretical account of stigma, Goffman (1963) observed that the primary challenge in social interactions faced by persons with a concealable stigma is to control who knows about their stigmatized status. He referred to persons with a concealable stigma as the *discreditable* to highlight the importance of such information management. As the term discreditable suggests, when one's stigma is revealed to others, it often carries negative consequences, ranging from social stereotypes inaccurately applied to oneself, to social ostracism and discrimination, to outright physical attack. Once an individual's stigma is revealed, according to Goffman (1963), he or she becomes one of the *discredited* and her or his primary task in social interaction shifts from managing personal information to attempting to influence how others use that information in forming impressions about the individual.

Gay men and lesbians frequently find this task complicated by the widespread perception that acknowledgment of their homosexual orientation is perceived as a highly intimate disclosure, unlike acknowledgment of heterosexuality. Self-disclosing gay people are likely to be regarded as inappropriately flaunting their sexuality, whereas heterosexuals' self-disclosures about their sexual orientation are usually considered not noteworthy because everyone is presumed to be heterosexual. This asymmetery creates difficulties in maintaining reciprocal levels of self-disclosure in social interactions between heterosexuals and homosexuals.

The foregoing discussion suggests that hiding their stigmatized status might be the safest strategy for gay men and lesbians. Successfully preventing others from learning about their stigma, however, requires considerable effort. *Passing* as a nonstigmatized person requires constant vigilance and a variety of strategies. These strategies include *discretion* (i.e., refraining from disclosing personal information to others), *concealment* (actively preventing others from acquiring information about oneself), and *fabrication* (deliberately providing false information about oneself to others; Zerubavel, 1982). Whichever strategy is used, passing requires the individual to lead a kind of double life (e.g., Ponse, 1976). It interferes with normal social interaction, creates a multitude of practical problems, and requires psychological as well as physical work.

Moreover, attempts to pass are not always successful. Lesbians and gay men often find that others have acquired information about their homosexuality from a third party, through astute observation, or simply by guessing. Even when they are able to pass, many gay people find the process personally objectionable for a variety of reasons. Consequently, they reveal their stigma to one or more other persons. Before elaborating further on these points, it is important to clarify the meaning of sexual orientation.

Sexual Orientation, Heterosexuality, and Homosexuality

Although heterosexual and homosexual behaviors alike have been common throughout human history, the ways in which cultures have made sense of these behaviors and the rules governing them have varied widely. For at least a century in the United States and Europe, human sexuality has been popularly understood in terms of a dichotomy between two types of people: those who are attracted to their same gender (*homosexuals*) and those who are attracted to the other gender (*heterosexuals*). (Individuals whose behavior crosses these categories have usually been labeled *bisexual* or their behavior has been explained as the product of situational or developmental factors such as a sex-segregated environment or an age-specific stage of sexual experimentation.)

This classification system differs from other possible ways of understanding sexuality in that its focus is the individual rather than the behavior. Instead of conceiving of people as capable of a wide range of sexual attractions and behaviors, the heterosexual-homosexual dichotomy creates two ideal *types* that, depending on the individual, correspond more or less to actual experience and behavior (for historical perspectives on the dichotomy, see Duberman, Vicinus, and Chauncey, 1989; Katz, 1983; for cross-cultural perspectives, see Herdt, 1984).

Sexual orientation is not simply about sex. Because sexual attraction and expression are important components of romantic relationships, sexual orientation is integrally linked to the close bonds that humans form with others to meet their personal needs for love, attachment, and intimacy. These bonds are not based only on specific sexual acts. They also encompass nonsexual physical affection, shared goals and values, mutual support, and ongoing commitment. In addition, one's sexual orientation is closely related to important personal identities, social roles, and community memberships. For heterosexuals, the identities and roles include those of *husband, wife, father,* and *mother.* Most heterosexuals experience their sexuality, their romantic and affectional relationships, and their social roles and community memberships based upon those relationships as a central component of who they are, that is, their sense of self or identity. For homosexual persons, being *gay* or *lesbian* is itself an important personal identity, one that is commonly associated with membership in a minority community, as elaborated below.

Although heterosexual and homosexual orientations alike encompass interpersonal relationships, personal identity, and community memberships, an asymmetry exists in U.S. society between the experiences of heterosexuals and gay people. The culture promotes an assumption of heterosexuality: Normal sexuality is equated with heterosexuality, and people are assumed to be heterosexual unless evidence is provided to the contrary (e.g., Herek, 1992; Hooker, 1965; Ponse, 1976). Consequently, heterosexuals need not disclose their sexual orientation for its own sake, but are free to do so incidentally to pursuing other goals. Gay men and lesbians, in contrast, routinely face negative social sanctions if their sexual orientation becomes public knowledge, with the consequence that disclosing it often becomes an important act of self-affirmation as well as a vehicle for meeting other needs. This asymmetry is briefly explored below.

Heterosexuality and Normalcy

Society's institutions and customs routinely elicit and convey information about individuals' heterosexuality. Advertisements and other messages in mass media explicitly convey the assumption that the audience consists of heterosexuals, many of whom are preoccupied with meeting, marrying, living with, or having sexual relations with someone of the other gender. Employers, schools, hospitals, and government institutions often request information about one's marital status, spouse, and children. People routinely are publicly identified as part of a heterosexual relationship, whether as a fiancé, spouse, or widow. Wedding rituals and anniversaries are important family and community events.

Patterns of normal social interaction also reflect the heterosexual assumption. Heterosexuals are (correctly) assumed to be heterosexual without ever explicitly revealing their sexual orientation to others; they need not come out of the closet. Nonetheless, most heterosexuals regularly make statements and provide information to others about their relationship status, attractions, and even their problems with establishing or maintaining heterosexual relationships. Wearing a wedding ring or displaying a photograph of one's spouse or (opposite-sex) romantic partner, for example, publicly identifies oneself as heterosexual.

Such affirmations of heterosexuality, however, are not commonly interpreted as statements about private sexual conduct. Rather, they identify an individual as occupying a particular role in society. These roles—husband or wife, father or mother—are largely *desexualized* (Herek, 1992). That is, they are interpreted by others primarily as indicators of social duties and behaviors; they are not perceived to be associated primarily or exclusively with sexual behaviors, even though they recognize private sexual conduct (and, in the case of marriage, legitimize such conduct).

When a man says that he is married or a woman says that she is a mother, for example, the recipient of this information could make assumptions about the individual's private sexual behavior with a high likelihood of being accurate. Presumably, the married man has at some time engaged in heterosexual intercourse with his wife (although such a statement does not reveal the frequency of such intercourse), and the mother can be presumed to have engaged in heterosexual intercourse at least once (although she might have become a parent through artificial insemination or adoption). Yet marriage and parenthood are not usually construed primarily in sexual terms. Even at the time of a wedding—when assumptions about sexual conduct are perhaps most explicit—sex is widely understood to be merely one part of a larger picture. Friends and relatives may expect that newlyweds will engage in sexual intercourse (and some wedding rituals include serious or joking references to this fact), but most do not regard the marriage in exclusively or primarily sexual terms. Indeed, advice to the newly married often stresses the many responsibilities and obligations associated with married status, rather than romance and sex (see, e.g., Slater, 1963).

Because of the desexualized nature of heterosexual social roles, disclosures that identify one as heterosexual are not perceived as an inappropriate communication of information about private sexual conduct. Thus, the act of referring to one's heterosexual spouse in conversation, or of introducing that spouse to one's coworkers, is not regarded as a flaunting of one's sexuality. Family members or friends may approve or disapprove of the spouse's character, physical appearance, race, religion, occupation, or social class. They may be happy that their friend or relative has settled down, or may feel that he or she should have waited longer before marrying. They may speculate about whether the couple's relationship is likely to endure. That a man's spouse is a woman, or that a woman's spouse is a man, does not often elicit surprise or comment. A heterosexual orientation is unremarkable, usually unproblematic, and taken for granted.

Homosexuality, Invisibility, and Stigma

By contrast, homosexuality is stigmatized in the United States. Historically, *the homosexual* has been defined as a counterpart to *the normal person*: People identified as homosexual have been regarded as abnormal and deviant, and have accordingly been stigmatized as sinners, criminals, and psychopaths (see chapter 9 regarding different constructions of homosexuality). Stigma persists to the present day (for a review, see Herek, 1995). Opinion surveys since the 1970s have consistently shown that roughly two-thirds of adults in the United States condemn homosexuality or homosexual behavior as morally wrong or a sin (Herek, forthcoming). Only a plurality of Americans feel that homosexual relations between consenting adults should be legal (Herek, forthcoming). In addition, more than half of heterosexual adults feel that homosexual relations—whether between women or men—are disgusting, and about three-fourths regard homosexuality as unnatural (Herek and Capitanio, 1996). Public revelation that one is a homosexual can have serious negative consequences, including personal rejection and isolation, employment discrimination, loss of child custody, harassment, and violence (Badgett, 1995; Berrill, 1992; Herek, 1995; Levine, 1979a; Levine and Leonard, 1984; Patterson, 1992). Heterosexuality not only remains the statistical norm, but is also the only form of sexuality that society regards as natural and legitimate.

Like heterosexuals, most individuals with a homosexual identity experience their sexual orientation as a core part of the self. Unlike heterosexuals, however, most lesbians and gay men also experience their sexual orientation as problematic—to at least some extent—for several reasons. First, society's assumption of heterosexuality means that most gay people were raised with the expectation that they would be heterosexual. Not conforming to this expectation, they had to discover and construct their homosexual identity against a backdrop of societal disapproval, usually without access to parental or familial support or guidance (Herdt, 1989; Martin, 1982; Savin-Williams, 1994). Because most people internalize society's negative attitudes toward homosexuality, the discovery that oneself is gay often involves overcoming denial and then integrating one's homosexuality into the rest of one's identity in a positive way; this is the process of coming out to oneself (Malyon, 1982; Stein and Cohen, 1984).

Second, like heterosexuality, a homosexual orientation is closely related to important personal identities, social roles, and community memberships. But because homosexuality has historically been defined as deviant and abnormal, homosexuals' identities and roles have been of an oppositional nature; that is, they represent the viewpoint of an outsider. Defining oneself personally and socially as *gay* or *lesbian*—or, more recently, *queer*—provides entry to alternative communities that have developed in the United States and elsewhere (Levine, 1979b; Murray, 1979; Warren, 1974), but frequently prevents one from participating in "normal" community activities.

Third, whereas the public roles that assert one's heterosexuality (e.g., husband, wife) are desexualized, the roles associated with homosexuality are sexualized. Heterosexual relationships are generally understood as involving many components, but same-sex relationships are widely perceived only in sexual terms, even though they are very similar to heterosexual relationships in that they are primarily about love, affection, and commitment (Blumstein and Schwartz, 1983; Kurdek, 1994; Kurdek and Schmitt, 1986; Peplau, 1991; Peplau and Cochran, 1990). Consequently, whereas a man and woman holding hands in a public setting are likely to be regarded fondly ("All the world loves a lover"), a similar public expression of affection between two men or two women is usually perceived as an inappropriate flaunting of sexuality.[1]

Despite these problems, a homosexual identity forms and develops through the same process as do other aspects of identity and the self: social interaction (Erikson, 1963; Mead, 1934). By stating "I am gay," "I am a lesbian," or "I am a homosexual" (or making such assertions through symbolic speech or other conduct), an individual affirms her or his identity and integrates it with other facets of the self. This process of affirmation and integration is generally recognized as an important component of identity formation and psychological health (Malyon, 1982; Stein and Cohen, 1984).

Thus, stigmatization of homosexuality creates a dilemma for lesbians and gay men. If they allow themselves to be perceived as heterosexual (or work to pass as heterosexual), they must lead a double life that requires considerable effort and carries psychological costs. If, however, they identify themselves as homosexual (or allow others to learn of their homosexuality), they are likely to be perceived as inappropriately flaunting their sexuality and they risk ostracism, discrimination, and even physical violence. Despite the risks, many gay men and lesbians today choose to come out to at least some heterosexuals. The sexual orientation of others is often revealed without their consent. Some negative and positive consequences of being out of the closet are discussed in the next section.

Coming Out: Disclosure of Homosexuality

Although some gay individuals (including some active-duty military personnel) have disclosed their sexual orientation by going public—e.g., by appearing on national television or disclosing their homosexuality in a newspaper report—survey research conducted with national probability samples suggests that this pattern is not common. In a 1991 national telephone survey of attitudes and opinions that I conducted with John Capitanio, approximately 45 percent of the heterosexual respondents who knew at least one gay person reported that they first learned about the individual's sexual orientation directly from that person herself or himself (Herek and Capitanio, 1996). Another 16 percent initially learned about it through a third party or guessed, but subsequently discussed it directly with the gay or lesbian friend or relative.

When we examined the types of relationships in which direct disclosure was made, we found that such disclosure almost always occurred between close friends and immediate family members, rarely occurred between distant relatives, and occurred slightly more than half the time between acquaintances or casual friends. Because this study was conducted with a nationally representative sample, we can conclude that approximately 61 percent of adult heterosexuals in the United States who know gay men or lesbians were told directly by at least one gay friend or relative about his or her homosexuality, and that such revelations occur more often in close relationships than in distant relationships.

Some revelations, of course, are out of the person's control; information about her or his homosexuality is circulated by a third party without the individual's consent. In recent years, such third-party disclosures have been referred to as "outing" (Gross, 1993; Johansson and Percy, 1994). In 32 percent of the relationships with a gay person reported by heterosexuals in the national sample (note that respondents could report more than one relationship), the respondent initially learned that the friend, relative, or acquaintance was homosexual through a third party. In another 30 percent of the relationships, the respondent initially guessed that the friend, relative, or acquaintance was gay.

In summary, the majority of heterosexuals who know gay people have been told directly by at least one person that he or she is gay. At the same time, gay people often are outed involuntarily or their sexual orientation is guessed by a heterosexual. Nevertheless, some gay men and lesbians keep their sexual orientation hidden from all or most of their social circles. In a 1989 national telephone survey of four hundred lesbians and gay men, for example, between 23 percent and 40 percent of the respondents (depending upon geographical region) had *not* told their family of their sexual orientation, and between 37 percent and 59 percent had *not* told their coworkers ("Results of poll," 1989).

The Effects of Being Out on Social Perceptions

As noted previously, homosexuality's stigmatized status means that people who are identified as gay or lesbian are likely to encounter differential treatment by others, including ostracism, discrimination, and violence. Being identified as homosexual also has subtle effects on the ways in which gay men and lesbians are perceived by heterosexuals. Once a person is known to be homosexual, others regard that fact as the most (or one of the most) important pieces of information they possess about her or him. In other words, homosexuality represents a master status (Becker, 1963). Knowledge of one's homosexuality colors all other information about the individual, even information that is totally unrelated to sexual orientation. Consequently, once a man or woman is labeled by others as a homosexual, all of her or his actions—regardless of whether they are related to sexual orientation—are likely to be interpreted in light of her or his sexual orientation. The master status of homosexuality has at least three important consequences.

First, gay-identified people are regarded by heterosexuals primarily in sexual terms. This sexualization of the individual is evident in the DoD's equation of disclosing that one is homosexual with acknowledgement that one has engaged in or intends to engage in homosexual behavior. Such assumptions are not necessarily accurate, however. A public statement about one's psychological identity and community membership may not reveal a great deal about one's private sexual behavior. A heterosexual who has never had sexual contact with another person (e.g., as a consequence of choosing to wait until marriage, taking a vow of celibacy, or lacking a suitable partner) or has not had sex for a long time (e.g., as a consequence of choice, aging, loss of a spouse, lack of a partner) is nevertheless a heterosexual. By the same logic, an individual may self-identify as homosexual, gay, or lesbian and yet not be engaging in sexual acts with others for a variety of reasons.

Second, gay-identified individuals are likely to experience problems establishing a satisfactorily reciprocal level of intimacy in day-to-day social relations. Using Archer's (1980) previously described criteria, revealing information about one's homosexuality is likely to be perceived as a highly intimate self-disclosure. That one is gay or lesbian is a characteristic that invites broad generalizations, is distinctive and not readily evident in normal social interaction, and whose disclosure can be an affect-laden event. Revealing that one is heterosexual, by contrast, is not regarded as intimate self-disclosure. Indeed, as noted above, revelation about one's heterosexual relationships or one's marital status is routine in casual interactions with strangers. Derlega, Harris, and Chaiken (1973), for example, found that a woman's disclosure of being caught by her mother in a sexual encounter was judged to be more intimate when the encounter was described as being with another woman than with a man.

A third important consequence of coming out is that popular stereotypes about homosexuals are applied to oneself (see chapter 9). A *stereotype* is a fixed belief that all or most members of a particular group share a characteristic that is unrelated to their group membership. Examples of widespread and enduring stereotypes are the beliefs that blacks are lazy and Jews are greedy. Some stereotypes of gay men and lesbians are also commonly applied to other disliked minority groups in this and other cultures. These include the stereotypes that members of the minority are hypersexual; a threat to society's most vulnerable members (e.g., children); secretive, clannish, and untrustworthy; and physically or mentally sick (Adam, 1978; Gilman, 1985). Other stereotypes are more specific to homosexuality, such as the beliefs that gay men are effeminate and lesbians are masculine (e.g., Kite and Deaux, 1987).

When people hold stereotypes about the members of a group, they tend to perceive and remember information about the group in a way that is consistent with their stereotypes. Heterosexuals who hold stereotypes about lesbians and gay men are more likely than others to engage in *selective perception* and *selective recall*. That is, they tend selectively to notice behaviors and characteristics that fit with their preconceived beliefs about gay men or lesbians, while failing to notice behaviors and characteristics that are inconsistent with those beliefs (e.g., Gross, Green, Storck, and Vanyur, 1980; Gurwitz and Marcus, 1978). And when they are trying to remember information about a gay person, their recollections and guesses about that individual tend to fit with their preconceived beliefs (e.g., Bellezza and Bower, 1981; Snyder and Uranowitz, 1978).

Stereotypical thinking is difficult to overcome, even for people who have consciously decided that they do not wish to be prejudiced. Although the latter are likely to experience guilt, discomfort, or other negative feelings when they realize that they have been thinking stereotypically about a particular group, they do it nevertheless (Devine and Monteith, 1993). Stereotypical thinking is resistant to change for several reasons. The use of stereotypes appears to be fairly automatic when thinking about members of a social out-group. In other words, using stereotypes is like a habit. To break the stereotype habit, one must break out of the automatic mental processes that one usually uses, and consciously take control of one's thinking. Such a change requires cognitive effort. It also requires acquisition of skills—one must learn new ways of (non-stereotypical) thinking (Devine and Monteith, 1993). Another reason that stereotypical thinking is difficult to overcome is that people tend to use whatever information is most accessible to them when they are making judgments and decisions (Tversky and Kahneman, 1973). Stereotypical beliefs often represent the most readily available information about the members of a social out-group. Finally, stereotypes persist because they tend to be reinforced by other members of one's own group. Someone who expresses a stereotypical belief about an out-group is likely to be rewarded by members of the in-group (e.g., in the form of acceptance or liking), whereas someone who expresses a counterstereotypical belief may be punished (in the form of disagreement, discounting, ridicule, or even rejection and ostracism).

Why Do Lesbians and Gay Men Come Out?

Given the prevalence of stigma and the negative consequences of being labeled a homosexual, it might be asked why lesbians and gay men ever voluntarily reveal their sexual orientation to anyone. At least three broad categories of reasons can be identified.

Improving interpersonal relationships. Withholding information about oneself from friends, coworkers, and acquaintances often disrupts social relationships—or hinders their development—and arouses suspicions about an individual's private life. As noted above, disclosure of information about oneself is an important component of forming and maintaining interpersonal relationships, with more extensive and intimate disclosures characterizing closer relationships. Because sexual orientation is so central to personal identity, keeping it a secret from another person necessarily requires withholding a substantial amount of information about oneself. This information is central to many of the topics that are commonly discussed by people in a close relationship. Examples include the joys and stresses of one's romantic relationships or search for such relationships, feelings of fulfillment or loneliness, and mundane or momentous experiences with one's partner. In most such discussions, the gender of one's partner (and, consequently, information about one's sexual orientation) is revealed simply through the accurate use of masculine or feminine pronouns. Thus, when one individual conceals his or her sexual orientation from another, the two cannot have an honest discussion of such matters. As a result, spontaneity and personal disclosure are necessarily limited, which inevitably impoverishes the relationship.

Keeping one's sexual orientation a secret also creates a variety of practical problems (e.g., ensuring that one's heterosexual acquaintances do not see one entering a gay club or church, or do not learn about the gender of one's lover) and ethical problems (e.g., lying and deception; see Plummer, 1975). Some gay people disclose to others as a way of eliminating or reducing these problems. Anticipating that others will find out anyway, some gay people disclose as a way of exercising some degree of control over others' perceptions (e.g., Davies, 1992; for examples of similar types of disclosure by members of other stigmatized groups, see Miall, 1989, and Schneider and Conrad, 1980). In either case, such disclosures represent an attempt to exercise control over the way in which another person learns of one's stigma when such revelation is inevitable, and to frame that information in a positive light. Thus, coming out makes the gay person's life simpler and makes possible honest relationships with others (Wells and Kline, 1987).

Enhancing one's mental and physical health. Research on stigma has documented the use of self-disclosure as a strategy for relieving the stress associated with concealment of one's stigma while enhancing one's self-esteem and overcoming the negative psychological effects of stigmatization (Schneider and Conrad, 1980). Such *therapeutic disclosure* usually requires an audience that is supportive, encouraging, empathetic, and nonjudgmental (Schneider and Conrad, 1980; but see Herman, 1993). Hence, it is most likely to occur with immediate family members or individuals who are considered close friends (e.g., Miall, 1989). By disclosing to such individuals, the stigmatized person can reduce or eliminate the negative feelings about himself or herself that often accompany secrecy and isolation, while developing a new, shared definition of his or her stigmatized attribute as normal and ordinary (Schneider and Conrad, 1980).

Gay people often disclose to others as a strategy for promoting their own well-being. As mentioned earlier, lesbians and gay men have been found to manifest better mental health to the extent that they feel positively about their sexual orientation and have integrated it into their lives through coming out and participating in the gay community (Bell and Weinberg, 1978; Hammersmith and Weinberg, 1973; Herek and Glunt, 1995; Leserman et al., 1994). In contrast, closeted gay women and men may experience a painful discrepancy between their public and private identities (Humphreys, 1972; see generally Goffman, 1963; Jones et al., 1984; for a discussion of AIDS stigma and passing, see Herek, 1990). They may feel inauthentic or that they are living a lie (Jones et al., 1984). Although they may not face direct prejudice against themselves, they face unwitting acceptance of themselves by individuals who are prejudiced against homosexuals (Goffman, 1963). Passing can also create considerable strain for lesbian and gay male couples, who must hide or deny their relationship to family and friends. This denial can create strains in the relationship and, when it prevents the partners from receiving adequate social support, may have a deleterious effect on psychological adjustment (Kurdek, 1988; Murphy, 1989).

Coming out may also promote physical health. Psychologists have long hypothesized that hiding or concealing significant aspects of the self can have negative effects on physical health, whereas disclosure of such information to others can have positive health consequences (e.g., Jourard, 1971). Empirical research has generally supported these hypotheses. Larson and Chastain (1990), for example, found that survey respondents high in self-concealment manifested significantly more bodily symptoms, depression, and anxiety than did respondents who were low in self-concealment. The negative health correlates of self-concealment appear to be independent of an individual's degree of social support (Larson and Chastain, 1990; Pennebaker and O'Heeron, 1984). Recent empirical research points to the physiological mechanisms underlying such relation-ships, indicating that ongoing inhibition of behavior—as is involved in active deception or concealment—requires physical effort and is accom-panied by short-term physiological changes, such as increased electroder-mal activity (Fowles, 1980). Pennebaker and Chew (1985), for example, observed an increase in skin conductance levels when experimental sub-jects (following the experimenter's instructions) actively deceived another individual.

Long-term behavioral inhibition may lead to stress-related disease (Pennebaker and Susman, 1988) and, conversely, disclosure of previously concealed personal information appears to be associated with better physical health. Pennebaker, Kiecolt-Glaser, and Glaser (1988), for ex-ample, found that individuals who wrote a series of essays in which they disclosed highly personal information about upsetting experiences subse-quently displayed better immune functioning, lower blood pressure, fewer medical visits, and less subjective distress than did members of a control group who wrote essays about trivial topics. Esterling, Antoni, Fletcher, Margulies, and Schneiderman (1994) observed lower levels of Epstein-Barr Virus (EBV) antibody titers (indicating better immune sys-tem functioning) among experimental subjects who verbally disclosed personal information about a stressful event, compared to subjects who disclosed such information in a written essay. The latter group, in turn, displayed lower EBV antibody titers than subjects in a control group (Es-terling et al., 1994).

In summary, coming out appears to be associated with enhanced men-tal health. In addition, although empirical research has not directly as-sessed whether deceiving others about one's sexual orientation can lead to physical health problems, such a conclusion is consistent with existing research.

Changing society's attitudes. As Goffman (1963) noted, some stigmatized individuals devote considerable energy and resources to self-disclosure in order to change societal attitudes and to help others who share their stigma (see also Anspach, 1979; Kitsuse, 1980). Some of them *go public,* that is, make their status a matter of public record through, for example, a speech, media interview, or legal proceeding (Lee, 1977).

Like members of other stigmatized groups (see, e.g., Schneider and Conrad, 1980, and Gussow and Tracy, 1968), gay people often come out to others to educate them about what it means to be gay, and to affect their actions toward gay people as a group. Indeed, many gay men and lesbians regard coming out as a political act that is a necessary prelude to changing society's treatment of them (Humphreys, 1972; Kitsuse, 1980). Perhaps the most noted political leader to advocate this strategy was Harvey Milk, San Francisco's first openly gay supervisor, who was assassinated in 1978. For example, in a message that he had recorded to be played in the event of his death, Milk expressed the belief that coming out would eliminate prejudice: "I would like to see every gay lawyer, every gay architect come out, stand up and let the world know. That would do more to end prejudice overnight than anybody could imagine" (Shilts, 1982, p. 374).

Social science theory and data suggest that coming out is indeed likely to have a positive effect on heterosexuals' attitudes toward gay people as a group. This prediction is based on the *contact hypothesis*, which states that intergroup hostility and prejudice can be reduced by personal contact between majority and minority groups in the pursuit of common goals (Allport, 1954; see also chapter 9). A large body of empirical data (e.g., Amir, 1976; Brewer and Miller, 1984; Stephan, 1985) indicates that intergroup contact can indeed change attitudes, provided that the contact meets the conditions originally specified by Allport (1954); for example, the interacting individuals should share equal status and the two groups should share superordinate goals.

Most empirical research using the contact hypothesis has focused on interracial and interethnic prejudice (for reviews, see Amir, 1976; Brewer and Miller, 1984; Stephan, 1985). Social psychologists recognize, however, that common psychological processes underlie all forms of intergroup prejudice, regardless of the specific out-group involved. The same theories and methods have been applied to understanding heterosexuals' antigay attitudes as have been used for, say, whites' antiblack attitudes (for examples, see Herek, 1987a, 1987b). Consequently, it would be reasonable to assume that the contact hypothesis is applicable to the case of heterosexuals' attitudes toward lesbians and gay men, even in the absence of data.

Supporting data, however, are indeed available. Survey research conducted with nationally representative probability samples (Herek and Capitanio, 1995, 1996; Herek and Glunt, 1993; Schneider and Lewis, 1984) and with nonrepresentative convenience samples (Doran and Yerkes, 1995; Gentry, 1987; Herek, 1988; Millham, San Miguel, and Kellogg, 1976; Weis and Dain, 1979) has consistently shown that heterosexuals who report personal contact with gay men or lesbians express significantly more favorable attitudes toward gay people as a group than do heterosexuals who lack contact experiences. In a 1988 national telephone survey of 937 adult U.S. residents, for example, Eric Glunt and I asked respondents, "Have any of your female or male friends, relatives, or close acquaintances let you know that they were homosexual?" We found that individuals who responded affirmatively had significantly lower scores on a measure of prejudice against gay men (Herek and Glunt, 1993). Furthermore, we observed that contact was associated with less prejudice regardless of a respondent's demographic characteristics (including gender, age, educational background, level of religiosity, marital status, number of children, and geographic region). We also found that contact was the best statistical predictor of respondents' attitudes toward gay men; that is, the contact variable explained a greater proportion of variance in attitudes than any other demographic or social variable that we assessed.

I replicated and expanded upon these findings in my previously mentioned research with John Capitanio (Herek and Capitanio, 1996). In a 1990-1991 national telephone survey of 538 adult heterosexuals, we asked respondents if they had "any male or female friends, relatives, or close acquaintances who are gay or homosexual." Respondents who reported contact experiences with at least one gay person (roughly one-third of the sample) manifested significantly more favorable attitudes toward gay men compared to respondents without contact experiences. Consistent with my study with Eric Glunt, we found that this pattern held across most demographic and social groups, and that contact was the most powerful predictor of attitudes.

In a follow-up survey one year later, we asked the same respondents about their attitudes toward lesbians and found the same patterns. In both surveys, we also asked questions about the nature of respondents' contact experiences: how many lesbians or gay men they knew, the type of relationship they had, and how they learned that a friend or relative was gay. Consistent with the contact hypothesis, we found that having a close relationship with a gay individual (e.g., a close friend or a member of one's immediate family) was associated with more favorable attitudes toward gay people generally than was having a more distant relationship (e.g., an acquaintance or distant relative). We also found that contact exerted an additive effect on attitudes: Respondents who knew three gay people generally had more favorable attitudes than those who knew two, who had more favorable attitudes than those who knew one.

The few available attitude studies conducted with military samples indicate that the relationship between contact experiences and favorable attitudes observed among civilians also holds for military personnel. Naval hospital personnel who reported having more than one gay friend manifested significantly less negative attitudes toward gay people generally (Doran and Yerkes, 1995), and Army personnel who believed that a gay man or lesbian was serving in their unit were more willing than others to allow homosexuals to serve in the military (Miller, 1993, cited in National Defense Research Institute, 1993, ch. 7). In another survey of Army personnel conducted by Miller (1994), having gay friends was a significant predictor of opposition to the military's ban on homosexual personnel.

Yet another finding from my national survey (Herek and Capitanio, 1996) concerned the effects of self-disclosure on others' attitudes. We found that respondents who had been told directly by a gay friend or relative about the latter's homosexuality had more favorable attitudes toward gay people as a group, compared to respondents who had guessed about a friend or relative's homosexuality, or had been told by a third party. This effect also appeared to be additive: Respondents' attitudes tended to be more favorable to the extent that they had been the recipient of self-disclosures from more gay or lesbian individuals (Herek and Capitanio, 1996).

Of course, correlational data do not indicate causality. But our analyses of the relationships between reports of contact in the first survey (referred to here as Wave 1) and the same respondents' attitudes one year later in the follow-up survey (Wave 2) indicated that heterosexuals who knew a gay man or lesbian in 1990-1991 were likely to develop more positive attitudes toward gay people as a group in the following year. Wave 1 contact explained a significant amount of variance in Wave 2 attitudes, even when Wave 1 attitudes were statistically controlled. At the same time, we also observed that heterosexuals who reported favorable attitudes toward gay men and lesbians in 1990-1991 were more likely than other respondents to experience contact with a gay person in the subsequent year. That is, Wave 1 attitudes predicted reports of contact at Wave 2, even controlling statistically for Wave 1 contact, probably because lesbians and gay men tend to reveal their sexual orientation to heterosexuals who they expect to react favorably (see also Wells and Kline, 1987). In summary, then, not only does contact with gay people affect heterosexuals' attitudes, but a heterosexual person's attitudes (or gay men and lesbians' perceptions of them) probably affects the likelihood that she or he will knowingly experience contact with gay people.

What are the psychological processes through which contact experiences influence heterosexuals' attitudes? In a close relationship, a gay or lesbian individual's direct disclosure about her or his homosexuality can provide the heterosexual with the necessary information and motivation to restructure her or his attitudes toward gay people as a group. This seems most likely to occur when the gay man or lesbian carefully manages the disclosure process so that the heterosexual can receive information (e.g., about what it means to be gay, about the gay person's similarity to other gay people) in the context of a committed relationship. For example, the gay person may self-disclose in a series of gradual stages, frame the disclosure in a context of trust and caring, explain why she or he did not disclose earlier, answer the heterosexual person's questions, and reassure the heterosexual that her or his past positive feelings and favorable judgments about the gay friend or relative are still valid.

Such interactions may help the heterosexual to keep in mind the other person's homosexual identity while observing behaviors that are inconsistent with stereotypes about gay people. Such a juxtaposition can facilitate the rejection of those stereotypes while fostering attitude change. If this experience leads the heterosexual person to accept that the friend or relative is indeed representative of the larger community of gay people—in other words, the friend or relative is not regarded as an anomaly—the heterosexual is likely to experience cognitive dissonance. On the one hand, she or he has strong positive feelings toward the gay friend or relative; on the other hand, she or he probably has internalized society's negative attitudes toward homosexuality. If the dissonance is resolved in favor of the friend or relative—an outcome that is more likely when the gay person plays an active role in imparting information about her or his stigmatized status—the heterosexual's attitudes toward gay people as a group are likely to become more favorable. The probability of favorable attitudes resulting from contact appears to be greater to the extent that a heterosexual has contact with more than one lesbian or gay man. Knowing multiple members of a stigmatized group is probably more likely to foster recognition of that group's variability than is knowing only one group member (Wilder, 1978). Knowing multiple members of a group may also reduce the likelihood that their behavior can be discounted as atypical (Rothbart and John, 1985).

Summary. Gay men and lesbians have a variety of reasons for disclosing their sexual orientation to others. Coming out affirms a core component of one's identity and facilitates the integration of one's homosexual identity with other aspects of the self. It permits honesty and openness in personal relationships with others, thereby enhancing and maintaining those relationships and creating a relational context in which other kinds of self-disclosure can occur. It permits the individual to feel authentic and to enjoy enhanced social and psychological functioning, while also possibly reducing stress and psychogenic symptoms. And it represents a political act through which an individual can attempt to change societal attitudes. Conversely, the negative consequences of staying in the closet include feelings of inauthenticity, impaired social relationships and interactions, increased strain on one's intimate relationships, and psychological and physical distress.

Conclusions and Implications for the Military

The foregoing discussion has several implications for military policy. First, it demonstrates that the current policy—by codifying society's norms about disclosure of sexual orientation—establishes different rights of expression for heterosexual and homosexual personnel. Heterosexual personnel are permitted to declare their sexual orientation publicly whereas homosexual personnel are not. A married heterosexual soldier, for example, can freely disclose information about her or his marital status, can publicly display a photograph or letter from the spouse, and can even publicly display affection for the spouse (e.g., holding hands, embracing, kissing)—all without negative sanctions. Furthermore, because heterosexual roles are desexualized, public affirmations that one is heterosexual are not construed as a presumption of sexual conduct, including conduct that is prohibited under the Uniform Code of Military Justice (e.g., oral sex).[2] Homosexual personnel, by contrast, are required to hide their sexual orientation publicly and, if their identity becomes known, are presumed to have engaged in illegal behavior.

Second, prohibiting gay men and lesbians from disclosing their sexual orientation does not simply mean that they are forbidden from discussing specific private sexual acts. Indeed, discussion of sexual behavior is a relatively minor component of public disclosure of one's sexual orientation. Current military policy has the effect of barring gay male and lesbian personnel from sharing a wide range of personal information with co-workers, friends, and acquaintances—information of the sort that is freely shared among heterosexuals.

Third, the prohibition on self-disclosure has important consequences for homosexuals. By requiring gay men and lesbians to hide significant portions of their lives, the policy imposes serious restrictions on their ability to interact socially. Whether gay people comply with the policy by using discretion, concealment, fabrication, or another strategy, they are disadvantaged—compared to heterosexuals—in establishing inter-personal relationships of the sort that contribute to social cohesion (see chapter 8) and opportunities for advancement. Furthermore, the ban on self-disclosure deprives gay men and lesbians of access to social support and may be deleterious to their long-term physical and psychological well-being.

Fourth, the policy prevents heterosexual personnel from interacting freely with openly gay men and women in the course of their duties. Ironi-cally, ongoing interpersonal contact would be likely to eliminate the prej-udicial attitudes that the DoD cites as the reason its policy is necessary. By allowing homosexuals the same rights of verbal self-disclosure currently permitted to heterosexuals, the military would create many of the condi-tions specified by social psychological research (e.g., institutional support for intergroup contact, shared goals) as likely to reduce interpersonal hos-tility and eliminate negative stereotypes.

Taken at face value, the maxim of "Don't Ask, Don't Tell, Don't Pur-sue" appears to promote a live-and-let-live atmosphere in which homo-sexual personnel are tolerated so long as they keep their sexual orienta-tion a private matter. As the foregoing discussion reveals, however, the policy places severe and sweeping strictures on gay people while pre-venting heterosexual personnel from experiencing the very types of social interactions that are most likely to eliminate antigay sentiment in the mili-tary's ranks. In a society in which homosexuality is stigmatized, to refrain from asking recruits about their sexual orientation is, perhaps, a positive step toward respecting the right of gay men and lesbians to retain control over information about their status. But in a society in which all adults are presumed to be heterosexual, to forbid gay people from telling others about their sexual orientation—and all aspects of their lives related to it—is to condemn them to invisibility and to sanction society's prejudice.

Notes

1. That same-gender relationships lack a desexualized public role comparable to that of husband or wife is evident in the terminology that gay men and lesbians have available for describing the individual with whom they share a committed relationship. Although *husband* and *wife* are legal terms, they also describe a complex set of relationships that, ideally, encompass the roles of lover, partner, and friend. Lacking legal spousal relationships, gay men and lesbians commonly use terms such as *lover*, *partner*, and *friend*, none of which conveys the complex set of meanings associated with *husband* or *wife*. Nor do those terms unambiguously describe the type of committed relationship signified by husband and wife. One's *partner* may be a business partner. One's *friend* may be simply an old school chum. One's *lover* may be a person with whom one is having a brief extramarital sexual affair. Each of these words creates confusion about the exact nature of the relationship; they describe only one part of it.

2. This assumption persists even though the majority of heterosexual adults in the United States—and, most likely, military personnel—have engaged in sexual behavior with their spouse that constitutes sodomy under the UCMJ, for example, oral sex (Laumann et al., 1994; see chapters 2 and 3).

References

Adam, B. D. (1978). *The survival of domination: Inferiorization and everyday life*. New York: Elsevier.

Allport, G. (1954). *The nature of prejudice*. New York: Addison-Wesley.

Altman, I., & Taylor, D. A. (1973). *Social penetration: The development of interpersonal relationships*. New York: Holt.

Amir, Y. (1976). The role of inter-group contact in change of prejudice and ethnic relations. In P. A. Katz (Ed.), *Towards the elimination of racism* (pp. 245–308). New York: Pergamon.

Anspach, R. R. (1979). From stigma to identity politics: Political activism among the physically disabled and former mental patients. *Social Science and Medicine, 13A*, 765–773.

Archer, R. L. (1980). Self-disclosure. In D. M. Wegner & R. R. Vallacher (Eds.), *The self in social psychology* (pp. 183–205). New York: Oxford University Press.

Badgett, M. V. L. (1995). The wage effects of sexual orientation discrimination. *Industrial and Labor Relations Review, 48*, 726–739.

Baxter, L. A. (1987). Self-disclosure and relationship engagement. In V. J. Derlega & J. H. Berg (Eds.), *Self-disclosure: Theory, research, and therapy* (pp. 155–174). New York: Plenum.

Becker, H. S. (1963). *Outsiders: Studies in the sociology of deviance*. New York: Free Press.

Bell, A. P., & Weinberg, M. S. (1978). *Homosexualities: A study of diversity among men and women.* New York: Simon & Schuster.

Bellezza, F. S., & Bower, G. H. (1981). Person stereotypes and memory for people. *Journal of Personality and Social Psychology, 41,* 856–865.

Berg, J. H., & Derlega, V. J. (1987). Themes in the study of self-disclosure. In V. J. Derlega & J. H. Berg (Eds.), *Self-disclosure: Theory, research, and therapy* (pp. 1–8). New York: Plenum.

Berrill, K. (1992). Anti-gay violence and victimization in the United States: An overview. In G. M. Herek & K. Berrill (Eds.), *Hate crimes: Confronting violence against lesbians and gay men* (pp. 19–45). Newbury Park, CA: Sage.

Blumstein, P., & Schwartz, P. (1983). *American couples.* New York: Morrow.

Brewer, M. B., & Miller, N. (1984). Beyond the contact hypothesis: Theoretical perspectives on desegregation. In N. Miller & M. B. Brewer (Eds.), *Groups in contact: The psychology of desegregation* (pp. 281–302). Orlando, FL: Academic Press.

Davies, P. (1992). The role of disclosure in coming out among gay men. In K. Plummer (Ed.), *Modern homosexualities: Fragments of lesbian and gay experience* (pp. 75–83). London: Routledge.

Davis, M. H., & Franzoi, S. L. (1986). Adolescent loneliness, self-disclosure, and private self-consciousness: A longitudinal investigation. *Journal of Personality and Social Psychology, 51,* 595–608.

Derlega, V. J., & Berg, J. H. (Eds.) (1987). *Self-disclosure: Theory, research, and therapy.* New York: Plenum.

Derlega, V. J., Harris, M. S., & Chaiken, A. L. (1973). Self-disclosure reciprocity, liking and the deviant. *Journal of Experimental Social Psychology, 9,* 277–284.

Devine, P. G., & Monteith, M. J. (1993). The role of discrepancy-associated affect in prejudice reduction. In D. M. Mackie & D. L. Hamilton (Eds.), *Affect, cognition, and stereotyping: Interactive processes in group perception* (pp. 317–344). San Diego, CA: Academic Press.

Doran, A. P., & Yerkes, S. A. (1995, August). *Attitudes toward gay men and lesbians in a Naval hospital sample.* Paper presented at the annual meeting of the American Psychological Association, New York.

Duberman, M. B., Vicinus, M., & Chauncey, G., Jr. (1989). *Hidden from history: Reclaiming the gay and lesbian past.* New York: New American Library.

Erikson, E. H. (1963). *Childhood and society* (2d ed.). New York: Norton.

Esterling, B. A., Antoni, M. H., Fletcher, M. A., Margulies, S., & Schneiderman, N. (1994). Emotional disclosure through writing or speaking modulates latent Epstein-Barr Virus antibody titers. *Journal of Consulting and Clinical Psychology, 62,* 130–140.

Fitzpatrick, M. A. (1987). Marriage and verbal intimacy. In V. J. Derlega & J. H. Berg (Eds.), *Self-disclosure: Theory, research, and therapy* (pp. 131–154). New York: Plenum.

Fowles, D. C. (1980). The three arousal model: Implications of Gray's two-factor learning theory for heart rate, electrodermal activity, and psychopathy. *Psychophysiology, 17*, 87–104.

Franzoi, S. L., & Davis, M. H. (1985). Adolescent self-disclosure and loneliness: Private self-consciousness and parental influences. *Journal of Personality and Social Psychology, 48*, 768–780.

Franzoi, S. L., Davis, M. H., & Young, R. D. (1985). The effects of private self-consciousness and perspective taking on satisfaction in close relationships. *Journal of Personality and Social Psychology, 48*, 1584–1594.

Gentry, C. S. (1987). Social distance regarding male and female homosexuals. *Journal of Social Psychology, 127*, 199–208.

Gilman, S. L. (1985). *Difference and pathology: Stereotypes of sexuality, race, and madness.* Ithaca, NY: Cornell University Press.

Goffman, E. (1963). *Stigma: Notes on the management of spoiled identity.* Englewood Cliffs, NJ: Prentice-Hall.

Gross, A. E., Green, S. K., Storck, J. T., & Vanyur, J. M. (1980). Disclosure of sexual orientation and impressions of male and female homosexuals. *Personality and Social Psychology Bulletin, 6*, 307–314.

Gross, L. (1993). *Contested closets: The politics and ethics of outing.* Minneapolis: University of Minnesota Press.

Gurwitz, S. B., & Marcus, M. (1978). Effects of anticipated interaction, sex, and homosexual stereotypes on first impressions. *Journal of Applied Social Psychology, 8*, 47–56.

Gussow, Z., & Tracy, G. S. (1968). Status, ideology, and adaptation to stigmatized illness: A study of leprosy. *Human Organization, 27*, 316–325.

Hammersmith, S. K., & Weinberg, M. S. (1973). Homosexual identity: Commitment, adjustment, and significant others. *Sociometry, 36* (1), 56–79.

Herdt, G. H. (1989). Gay and lesbian youth, emergent identities, and cultural scenes at home and abroad. *Journal of Homosexuality, 17* (1/2), 1–42.

———. (Ed.). (1984). *Ritualized homosexuality in Melanesia.* Berkeley, CA: University of California Press.

Herek, G. M. (1987a). Religion and prejudice: A comparison of racial and sexual attitudes. *Personality and Social Psychology Bulletin, 13* (1), 56–65.

———. (1987b). Can functions be measured? A new perspective on the functional approach to attitudes. *Social Psychology Quarterly, 50* (4), 285–303.

———. (1988). Heterosexuals' attitudes toward lesbians and gay men: Correlates and gender differences. *The Journal of Sex Research, 25*, 451–477.

———. (1990). Illness, stigma, and AIDS. In P. Costa and G. R. VandenBos (Eds.), *Psychological aspects of serious illness* (pp. 103–150). Washington, DC: American Psychological Association.

———. (1992). The social context of hate crimes: Notes on cultural heterosexism. In G. M. Herek & K. T. Berrill (Eds.), *Hate crimes: Confronting violence against lesbians and gay men* (pp. 89–104). Newbury Park, CA: Sage.

———. (1993). Sexual orientation and military service: A social science perspective. *American Psychologist, 48*, 538–547.

————. (1995). Psychological heterosexism in the United States. In A. R. D' Auge-
lli & C. J. Patterson (Eds.), *Lesbian, gay, and bisexual identities across the
lifespan: Psychological perspectives* (pp. 321–346). New York: Oxford University Press.

————. (forthcoming). The HIV epidemic and public attitudes toward lesbians
and gay men. In M. P. Levine, P. Nardi, & J. Gagnon (Eds.), *In changing times:
The impact of HIV/AIDS on gay men and lesbians.* Chicago: University of
Chicago Press.

Herek, G. M., & Capitanio, J. P. (1995). Black heterosexuals' attitudes toward
lesbians and gay men in the United States. *The Journal of Sex Research, 32,*
95–105.

————. (1996). "Some of my best friends": Intergroup contact, concealable
stigma, and heterosexuals' attitudes toward gay men and lesbians. *Personality
and Social Psychology Bulletin, 22,* 412–424.

Herek, G. M., & Glunt, E. K. (1993). Interpersonal contact and heterosexuals'
attitudes toward gay men: Results from a national survey. *The Journal of Sex
Research, 30,* 239–244.

————. (1995). Identity and community among gay and bisexual men in the AIDS
era: Preliminary findings from the Sacramento Men's Health Study. In G. M.
Herek & B. Greene (Eds.), *AIDS, identity, and community: The HIV epidemic
and lesbians and gay men* (pp. 55–84). Newbury Park, CA: Sage.

Herman, N. J. (1993). Return to sender: Reintegrative stigma management strategies of ex-psychiatric patients. *Journal of Contemporary Ethnography, 22,*
295–330.

Hooker, E. (1965). An empirical study of some relations between sexual patterns
and gender identity in male homosexuals. In J. Money (Ed.), *Sex research: New
developments* (pp. 24–52). New York: Holt, Rinehart, & Winston.

Humphreys, L. (1972). *Out of the closets: The sociology of homosexual liberation.* Englewood Cliffs, NJ: Prentice-Hall.

Johansson, W., & Percy, W. A. (1994). *Outing: Shattering the conspiracy of si-
lence.* New York: Harrington Park Press.

Jones, E. E., Farina, A., Hastorf, A. H., Markus, H., Miller, D. T., & Scott, R. A.
(1984). *Social stigma: The psychology of marked relationships.* New York:
W.H. Freeman.

Jourard, S.M. (1971). *The transparent self* (2d ed.). Princeton, NJ: Van Nostrand.

Katz, J. N. (1983). *Gay/lesbian almanac: A new documentary.* New York:
Harper & Row.

Kite, M. E., & Deaux, K. (1987). Gender belief systems: Homosexuality and the
implicit inversion theory. *Psychology of Women Quarterly, 11,* 83–96.

Kitsuse, J. I. (1980). Coming out all over: Deviants and the politics of social problems. *Social Problems, 28,* 1–13.

Kurdek, L. A. (1988). Perceived social support in gays and lesbians in cohabiting
relationships. *Journal of Personality and Social Psychology, 54,* 504–509.

Kurdek, L. A. (1994). The nature and correlates of relationship quality in gay, lesbian, and heterosexual cohabiting couples: A test of the individual difference, interdependence, and discrepancy models. In B. Greene & G. M. Herek (Eds.), *Lesbian and gay psychology: Theory, research, and clinical applications* (pp. 133–155). Newbury Park, CA: Sage.

Kurdek, L. A., & Schmitt, J. P. (1986). Relationship quality of partners in heterosexual married, heterosexual cohabiting, gay, and lesbian relationships. *Journal of Personality and Social Psychology, 51,* 711–720.

Larson, D. G., & Chastain, R. L. (1990). Self-concealment: Conceptualization, measurement, and health implications. *Journal of Social and Clinical Psychology, 9,* 439–455.

Laumann, E. O., Gagnon, J. H., Michael, R. T., & Michaels, S. (1994). *The social orgnization of sexuality: Sexual practices in the United States.* Chicago: University of Chicago Press.

Lee, J. A. (1977). Going public: A study in the sociology of homosexual liberation. *Journal of Homosexuality, 3* (1), 49–78.

Leserman, J., DiSantostefano, R., Perkins, D. O., & Evans, D. L. (1994). Gay identification and psychological health in HIV-positive and HIV-negative gay men. *Journal of Applied Social Psychology, 24,* 2193–2208.

Levine, M. P. (1979a). Employment discrimination against gay men. *International Review of Modern Sociology, 9* (5–7), 151–163.

———. (1979b). Gay ghetto. In M. P. Levine (Ed.), *Gay men: The sociology of male homosexuality* (pp. 182–204). New York: Harper & Row.

Levine, M. P., & Leonard, R. (1984). Discrimination against lesbians in the work force. *Signs, 9,* 700–710.

Malyon, A. K. (1982). Psychotherapeutic implications of internalized homophobia in gay men. *Journal of Homosexuality, 7* (2/3), 59–69.

Martin, A. D. (1982). Learning to hide: The socialization of the gay adolescent. In S. C. Feinstein, J. G. Looney, A. Z. Schwartzberg, & A. D. Sorosky (Eds.), *Adolescent psychiatry: Developmental and clinical studies* (Vol. 10, pp. 52–65). Chicago: University of Chicago Press.

Mead, G. H. (1934). *Mind, self, and society: From the standpoint of a social behaviorist.* Chicago: University of Chicago Press.

Miall, C. E. (1989). Authenticity and the disclosure of the information preserve: The case of adoptive parenthood. *Qualitative Sociology, 12,* 279–302.

Miller, L. L. (1994). Fighting for a just cause: Soldiers' views on gays in the military. In W. J. Scott & S. C. Stanley (Eds.), *Gays and lesbians in the military: Issues, concerns, and contrasts* (pp. 69–85). New York: Aldine de Gruyter.

Millham, J., San Miguel, C. L., & Kellogg, R. (1976). A factor-analytic conceptualization of attitudes toward male and female homosexuals. *Journal of Homosexuality, 2* (1), 3–10.

Murphy, B. C. (1989). Lesbian couples and their parents: The effects of perceived parental attitudes on the couple. *Journal of Counseling and Development, 68,* 46–51.

Murray, S. O. (1979). Institutional elaboration of a quasi-ethnic community. *International Review of Modern Sociology, 9,* 165–178.

National Defense Research Institute. (1993). *Sexual orientation and U.S. military personnel policy: Options and assessment.* Santa Monica, CA: RAND.

Patterson, C. J. (1992). Children of lesbian and gay parents. *Child Development, 63,* 1025–1042.

Pennebaker, J. W., & Chew, C. H. (1985). Behavioral inhibition and electrodermal activity during deception. *Journal of Personality and Social Psychology, 49,* 1427–1433.

Pennebaker, J. W., Kiecolt-Glaser, J. K., & Glaser, R. (1988). Disclosure of traumas and immune function: Health implications for psychotherapy. *Journal of Consulting and Clinical Psychology, 56,* 239–245.

Pennebaker, J. W., & O'Heeron, R. C. (1984). Confiding in others and illness rates among spouses of suicides and accidental death victims. *Journal of Abnormal Psychology, 93,* 473–476.

Pennebaker, J. W., & Susman, J. R. (1988). Disclosure of traumas and psychosomatic processes. *Social Science and Medicine, 26,* 327–332.

Peplau, L. A. (1991). Lesbian and gay relationships. In J. Gonsiorek & J. Weinrich (Eds.), *Homosexuality: Research findings for public policy* (pp. 177–196). Newbury Park, CA: Sage.

Peplau, L. A., & Cochran, S. D. (1990). A relationship perspective on homosexuality. In D. P. McWhirter, S. A. Sanders, & J. Reinisch (Eds.), *Homosexuality/ heterosexuality: Concepts of sexual orientation* (pp. 321–349). New York: Oxford University Press.

Plummer, K. (1975). *Sexual stigma: An interactionist account.* London: Routledge & Kegan Paul.

Ponse, B. (1976). Secrecy in the lesbian world. *Urban Life, 5,* 313–338.

Results of poll. (1989, June 6). *San Francisco Examiner,* p. A19.

Rothbart, M., & John, O. P. (1985). Social categorization and behavioral episodes: A cognitive analysis of the effects of intergroup contact. *Journal of Social Issues, 41* (3), 81–104.

Savin-Williams, R. C. (1994). Verbal and physical abuse as stressors in the lives of lesbian, gay male, and bisexual youths: Associations with school problems, running away, substance abuse, prostitution, and suicide. *Journal of Consulting and Clinical Psychology, 62,* 261–269.

Schneider, J. W., & Conrad, P. (1980). In the closet with illness: Epilepsy, stigma potential, and information control. *Social Problems, 28,* 32–44.

Schneider, W., & Lewis, I. A. (1984, February/March). The straight story on homosexuality and gay rights. *Public Opinion, 7,* 16–20, 59–60.

Shilts, R. (1982). *The mayor of Castro Street: The life and times of Harvey Milk.* New York: St. Martin's Press.

Slater, P. (1963). Social limitations on libidinal withdrawal. *American Sociological Review, 28,* 339–364.

Snyder, M., & Uranowitz, S. W. (1978). Reconstructing the past: Some cognitive consequences of person perception. *Journal of Personality and Social Psychology, 36,* 941–950.

Stein, T. S., & Cohen, C. J. (1984). Psychotherapy with gay men and lesbians: An examination of homophobia, coming out, and identity. In E. S. Hetrick & T. S. Stein (Eds.), *Innovations in psychotherapy with homosexuals* (pp. 60–73). Washington, DC: American Psychiatric Press.

Stephan, W. G. (1985). Intergroup relations. In G. Lindzey & E. Aronson (Eds.), *Handbook of social psychology* (Vol. 2, pp. 599–658). New York: Random House.

Stokes, J. P. (1987). The relation of loneliness and self-disclosure. In V. J. Derlega & J. H. Berg (Eds.), *Self-disclosure: Theory, research, and therapy* (pp. 175–201). New York: Plenum.

Tversky, A., & Kahneman, D. (1973). Availability: A heuristic for judging frequency and probability. *Cognitive Psychology, 5,* 207–232.

Warren, C. A. B. (1974). *Identity and community in the gay world.* New York: Wiley.

Weis, C. B., & Dain, R. N. (1979). Ego development and sex attitudes in heterosexual and homosexual men and women. *Archives of Sexual Behavior, 8,* 341–356.

Wells, J. W., & Kline, W. B. (1987). Self-disclosure of homosexual orientation. *Journal of Social Psychology, 127* (2), 191–197.

Wilder, D. A. (1978). Reduction of intergroup discrimination through individuation of the out-group. *Journal of Personality and Social Psychology, 36,* 1361–1374.

Zerubavel, E. (1982). Personal information and social life. *Symbolic Interaction, 5,* 97–109.

Sexual Modesty, the Etiquette of Disregard, and the Question of Gays and Lesbians in the Military

Lois Shawver

The tradition of segregating the toilet of men and women is being used to justify a ban on gays and lesbians in the military. Modesty requires that men and women dress and toilet separately, the argument goes, because undressing together would inevitably turn the situation into a sexual one. If this is true for heterosexuals, the argument continues, it must be true for gays and lesbians as well, hence gays and lesbians should not be allowed to serve in the military. They would inevitably be thrown into situations of undress and these privacy situations would likely become sexual. This is the "privacy argument" for the military ban on gays and lesbians, and it has found a central place in the thinking of those who objected to President Clinton's campaign promise to lift the ban.

In 1991 and 1992, although the ban was receiving public scrutiny, the privacy argument was stated separately by the chairman of the Joint Chiefs of Staff, General Colin Powell[1] and by D.C. Circuit Judge Oliver Gasch in a significant legal decision. Gasch's refusal to reinstate a gay man who had been discharged from the Navy illustrates the argument:

The embarrassment of being naked between the sexes is prevalent because sometimes the other is considered to be a sexual object. The quite rational assumption in the Navy is that with no one present who has a homosexual orientation, men and women alike can undress, sleep, bathe, and use the bathroom without fear or embarrassment that they are being viewed as sexual objects (*Steffan v. Cheney,* 1991).

This privacy argument is flawed. It is an argument from analogy built on a faulty premise. It mistakenly assumes that cross-gender undress is naturally erotic for heterosexuals, and then argues by analogy that self-gender undress must be equally so for homosexuals. What is erotic about undress for heterosexuals, however, is largely a matter of expectations. In real situations of undress, both heterosexuals and homosexuals can defuse sexual tensions and they ordinarily do so.

This cultural illusion in the United States, that cross-gender nakedness is inevitably erotic for heterosexuals, is based on several traditions. First, the United States has an elaborate modesty tradition. Gender segregation of toileting minimizes the number of situations in which people actually see members of the opposite sex undressed. Such minimal experience fosters unrealistic expectations about the experience of cross-gender undress.

Second, the United States has a practice of making highly eroticized images of partially clad people, especially women, evident in public places. Magazine stands are splashed with pictures of partially dressed women who have been carefully posed and photographed to enhance their sex appeal. And so while people in the United States do not often see images of old, heavy, unattractive, or awkwardly posed naked bodies, they frequently do see images of undress that have been made to appear as erotic as possible. Moreover, these images are presented impersonally as pictures that allow the viewer to look upon nakedness without the gaze being returned in a way that might be embarrassing. Such artificial images romanticize nakedness and foster the illusion that real situations of undress would be equally erotic.

When bodily exposure is highly unusual in real situations, exposure can acquire symbolic meaning that is culturally specific. Only in a culture that hides the ankle can an exposed ankle symbolize a romantic invitation. Exposure of the breast or thigh means something different in cultures with different traditions. These sexual symbols are treated as evidence within the culture that such exposures are naturally erotic, and such presumed evidence leads to myths that are made believable when voiced by authorities such as Gasch, Powell, and military sociologist Charles Moskos.

All these traditions converge to make it appear that the sexual overtones of cross-gender undress are natural and not culture bound. But despite folklore wisdom to the contrary, wherever nakedness is routine and not romanticized, both in the United States and in other countries, there is strong inclination for men and women to treat nakedness with an etiquette of disregard. What is potentially erotic, or embarrassing, is simply politely ignored. Any embarrassment and eroticism are contextually muted.

In contrast to the elaborate modesty ceremonies of the heterosexual culture in the United States, the homosexual culture minimizes modesty ceremony. Although heterosexuals do not often undress in cross-sex settings, gays and lesbians routinely frequent public lavatories, dressing rooms, and locker rooms, and share dormitories with heterosexuals without restriction. In these settings, they treat heterosexuals with an etiquette of disregard. Whereas the culture exaggerates the sexual potential of cross-gender undress, it ignores the sexual potential of homosexuals undressing with same-sex heterosexuals. The ability of homosexuals to be unnoticed in these situations of undress is testimony to their capacity to treat heterosexuals with an etiquette of disregard.

Those who make the privacy argument for the ban on gays and lesbians in the military have not taken into account the power of this etiquette to defuse sexual situations. The privacy argument wrongly presumes that people are unable to disregard the potential sexuality of situations. It assumes that modesty practices that hide nakedness are the only way to control sexual desire. The literature, however, suggests otherwise. This chapter will review relevant information about modesty and the etiquette of disregard.

The Nature of Sexual Modesty in the United States

Sexual modesty is the inclination to hide one's nakedness or bodily functions. It can be practiced either by hiding nakedness and toileting behind closed doors or by covering the body. In his classic 1901 article on modesty, Havelock Ellis defined this modesty inclination as a fear. He said, "Modesty . . . may be provisionally defined as an almost instinctive fear prompting to concealment" (1901/1942, p. 1).

In the United States, cross-gender modesty is cloaked in romantic ceremony. It is not merely a matter of hiding nakedness; it is the cultural ceremony of having zippers zipped and buttons buttoned even if the open garment does not expose naked parts of the body. Similarly, a modest woman does not appear on the street in her slip even though the slip might cover enough of her body to meet traditions. A modest man does not appear in public in his boxer shorts, although he might go running in gym shorts that expose as much or more of his body.

Modesty as a Cultivated Sensitivity

Modesty is a cultivated sensitivity that is passed on within the family culture (Parke and Sawin, 1979; Rosenfield et al., 1984; Sears, Maccoby, and Levin, 1957). Children are taught these modesty habits much as they are taught table manners, being scolded when they expose their bodies or underwear as they would be scolded if they eat their spaghetti with their fingers. Articles on how to teach modesty appear in popular parenting literature (e.g., Katz, 1989; Leonard, 1992; Segal and Segal, 1990). Rules are enacted and enforced (Rosenfield et al., 1984). Role modeling also appears involved. Retrospective questionnaire studies indicate that children imitate the modesty behaviors of parents (Lewis and Janda, 1988; Shelley, 1981) and of siblings and peers (Rosenfield et al., 1984). There is also research suggesting that girls are given more restrictive modesty training than boys (Maccoby and Jacklin, 1974), with little girls showing modesty behaviors beginning between ages four and six and little boys between five and eight. Subsequent to this training, modesty is well established for both genders by early adolescence (Parke and Sawin, 1979; Rosenfield et al., 1984).

The degree of modesty people feel varies across cultures. Although there has been a decline in cross-cultural research on modesty in the last fifteen years, there is strong evidence that North Americans are more modest than are people from most other Western countries. For example, in 1981 Ronald and Juliette Goldman asked 838 five- to fifteen-year-olds from Australia, England, Sweden, the United States, and Canada, "Suppose we all lived in a nice warm place or climate, should we need to wear clothes?" They found the North American subjects the most prudish and the Australians the least. Nevertheless, by thirteen years of age, most subjects (53 percent) answered that children in warm countries should wear clothes (Goldman and Goldman, 1981). Other authors have found North Americans to be more prudish than most people of European backgrounds (Kinsey, Pomeroy, and Martin, 1948; Sears, Maccoby, and Levin, 1957; Smith, 1979; Moore, 1984; Weinberg, 1964; Whiting and Child, 1953). On the other hand, during this same period, people from Muslim cultures were even more modest than people from English-speaking cultures (Antoun, 1968; Crawley, 1965; Fielding, 1942; Stephens, 1971).

Modesty as a Cultural Aphrodisiac

According to historians, in the late eighteenth and nineteenth centuries bodily modesty was thought to create sexual passion in men and to provide the key to sexual happiness (Rugoff, 1971; Walters, 1974). In his turn-of-the-century classic treatise on modesty, Ellis said, "The woman who is lacking in this kind of fear is lacking, also, in sexual attractiveness to the average man. . . . The immense importance of feminine modesty in creating masculine passion must be fairly obvious." He quoted passages from Casanova and others in support of his position (1901/1942, p. 1).

Many other Victorian writers made similar statements. For example, one book popular enough to be published through twenty-two editions stated:

> One of the first things that a mother seeks to instill into the mind of her little girl, is a feeling of shame which centers about the pelvic organs and their functions. This feeling, together with shyness, bashfulness, timidity, etc., develops a modesty which constitutes one of the chief, if not the greatest, of feminine charms. The mother is paving the way for her daughter's future happiness, for this commendable virtue not only acts as a shield and protection to the girl, but, by giving play to the imagination, provides for the happiness of her future lover (Malchow, 1923, pp. 60–61).

As this passage suggests, people in Victorian times were startlingly prudish by today's standards (Rugoff, 1971; Walters, 1974). Women wore their dresses to the floor. Exposing an ankle was deemed scandalous (Malchow, 1923). People in English-speaking countries began putting pinafores around tables and chairs to hide the legs (all legs were obscene), placing books by male and female authors on different shelves (Muir, 1983), and censoring the more sexually explicit words from Shakespeare and the Bible (Perin, 1969). All of this is a ceremony of modesty.

The Victorians, who lived in the cultural context of an exaggerated modesty ceremony, hypothesized that these traditions enriched their romantic imagery and enhanced their heterosexual passions. Current levels of modesty ceremony in the United States may perpetuate this tradition for heterosexuals in that men and women do not often see each other's bodies except when they have been dressed or posed to be alluring. Modesty ceremony serves to protect people from routinized exposure to undress with potential sex objects, which would teach them that it is possible to avoid erotic response in these situations.

Although the military's privacy argument assumes that modesty is needed to reduce sexual excitement, the experts on the practice of modesty, the Victorians, had a very different picture of the function of modesty. They hypothesized that modesty intensified sexual passion and served as a kind of aphrodisiac for men.

The Etiquette of Disregard

Research on the etiquette of disregard lends credence to the Victorian sexual hypothesis by providing evidence that without modesty ceremony, situations of undress do not turn into sexual situations. That is, when undress is routinized, people do not treat each other as sex objects as the privacy argument contends. People may feel an erotic flush when seeing a photograph of a partially clad model, but it is another matter entirely to be confronted by an undressed person who can recognize and return or reject a romantic gaze.

There are several settings in the United States in which toileting and undress are not hidden behind the cloak of sexual modesty. In these settings, sexual tensions are defused by a prevalent etiquette of disregard.

The etiquette of disregard is a polite habit of ignoring or disregarding that which might be embarrassing or have unintended sexual implications (Shawver, 1995). People use this etiquette in all sorts of situations whenever they believe it would be impolite to stare. It can serve to protect people's feelings about a deformity or a disability as well as to minimize embarrassment around nakedness and toileting. This etiquette routinely protects people from curiosity stares in public rest rooms, locker rooms, department store dressing rooms, saunas, and a variety of other private situations that are shared by people of the same or the opposite gender.

Although the term *etiquette of disregard* is new, there is a family of related concepts that are operationally defined in the research literature. These terms all reflect a cultural concern with an etiquette that requires a conformity to a standard of polite disregard in embarrassing or sexual situations. Gaze aversion (e.g., Abele, 1986; Kleinke, 1986) is one operational measure of the etiquette of disregard. Another is avoidance or denial of the embarrassing event (Cupach, Metts, and Hazelton, 1986; Edelmann, 1994). Still another is some form of social levity (Edelmann and

Iwawaki, 1987; Imahori and Cupach, 1994). There are many other measures of this etiquette specific to particular studies.

Research shows that gaze aversion is ingrained in American children
by the time they are of school age. Scheman and Lockard (1979) studied
childhood development of gaze aversion using adult experimenters who
stared at children in a shopping center. They found that although toddlers
did not avert their gaze, five- to nine-year-olds did. Asendorpf (1990)
reported similar findings. Gaze aversion is a common technique for minimizing embarrassment for both adolescents (Flaming and Morse, 1991)
and adults (Asendorpf, 1990; Edelmann, 1994; Edelmann and Iwawaki,
1987).

Kleinke (1986) reviewed studies that suggested that gaze aversion
could serve as a way of managing and diminishing the intimacy of situations. For example, Edelmann and Hampson (1981) had subjects participate in a question-and-answer session. Half the subjects were asked intimate questions whereas the control group had only nonintimate
questions. The subjects with intimate questions had decreased eye contact. Abele (1986) also found that subjects demonstrated more gaze aversion in interviews when they were asked highly intimate questions,
thereby replicating reports of earlier findings (Exline, Gray, and Schuette,
1965; Schulz and Barefoot, 1974). In summary, when adolescents or
adults in the United States find themselves in potentially embarrassing or
exposing situations, they are likely to feel a strong inclination to treat
the situation with an etiquette of disregard and look away.

"Avoidance" is another concept that sometimes operationalizes aspects of the etiquette of disregard. It might include changing the topic,
being silent, pretending that the embarrassing event did not take place,
or making any number of gestures that deny recognition to an embarrassment. Edelmann (1994) and Cupach, Metts, and Hazleton (1986)
reviewed several areas of embarrassment management research and listed
avoidance as a major technique of embarrassment management, especially for people from English-speaking countries.

The power of the etiquette of disregard to diminish the sexuality of a
situation is probably enhanced by the tendency of both men (Koukounas
and Over, 1992; O'Donohue and Geer, 1985) and women (Meuwissen
and Over, 1990; Zillman and Bryant, 1988) to acclimate to the presence
of a potentially erotic stimulus when the stimulus is no longer novel. People whose societies romanticize nakedness by restricting it to novel and
highly erotic contexts probably underestimate the ease with which the

erotic potential of communal undress can be overlooked when settings of undress are routinized in nonromantic contexts.

There is also evidence that subjects can control their level of sexual arousal. Mahoney and Strassberg (1991) showed heterosexual men explicitly sexual videos and instructed them on how much to allow themselves to be aroused by the experience. Measures of penile arousal indicated that subjects had a moderate ability to diminish their sexual response when so instructed. This study also found that sexual responses were more difficult to suppress when subjects were asked to focus on the details of the experience, indicating that gaze aversion and other forms of disregard can diminish sexual excitement. Other studies have demonstrated that sexual arousal is under many people's control (Dekker and Everaerd, 1988, 1989).

The research reviewed below, and a reflection on common experience, will show that in most situations of communal undress or toileting, people find themselves disregarding the potential eroticism of the situation.

Etiquette of Disregard in Other Cultures
With the Westernization of Japan in recent years, a number of researchers have compared the way the Japanese and the people of English-speaking cultures handle privacy situations.

Traditional Japanese culture often shocked Western travelers when they encountered Japan's general lack of modesty ceremony between the sexes. Before World War II, it was common for men and women in Japan to share public rest rooms without enclosed stalls. Men and women also typically shared public baths. American travelers were often amazed that the Japanese did not find these situations erotic (Downs, 1990).

Even today the Japanese are more comfortable with communal nakedness and toileting than are North Americans. Nevertheless, they are less comfortable than are Americans with the eroticized nakedness that splashes across American tabloids, theater, television and movie screens, and popular Japanese articles continue to describe the way the Japanese are scandalized by the American eroticizing of nakedness (Matsubara, 1994; Taylor and Baker, 1992). Even though the Japanese are more comfortable with nudity than Americans, the Japanese are more chaste than Americans (Shimazaki and Kaji, 1993). Whereas one-third of U.S. boys have had sex by age fifteen, only 6 percent of Japanese boys have done so by this age (Toufexis, 1993). And reports on the results of sexual education in the schools indicate that Japanese teachers are more reluctant than are U.S. teachers to discuss sexual issues openly (Kitazawa and Beerman, 1993; Shimazaki and Kaji, 1993).

Today, when there is evidence that Japan is gradually absorbing Western romanticized modesty traditions, there appears to be a simultaneous increase in sexual activity and interest. Asayama (1976) compared large survey samples of Japanese youths taken at intervals of several years (1952, 1960, and 1974), and found a marked increase in sexual awareness as well as an increase in sexual behaviors such as masturbation and kissing.

The Western tradition of romanticizing nakedness is also reflected in a greater need of English-speaking people to shift moods and deliberately conform to an etiquette of disregard to minimize eroticism in toileting situations. Research shows that although people from English-speaking nations generally have a higher base rate of gazing at others than do the Japanese (Argyle, 1975), the English-speaking people report more conscious aversion of their gaze in embarrassing situations than do the Japanese (Edelmann, 1990; Edelmann and Iwawaki, 1987).

The Japanese are not alone in having treated nakedness and toileting with an etiquette of disregard that minimizes the need for modesty ceremony and cross-gender hiding of the body. Seabrook and Burchill (1994) described how people of many countries in Africa and Southeast Asia have long histories of treating nakedness as routine and not erotic, thereby minimizing sexual allure in public settings with an etiquette of disregard. "Observations of daily life among naked tribes indicate that the sight of genitalia in a nude society is not in itself erotic" (Goodson, 1991, p. 191). Considerable work in demystifying the public about the universality of the Western romanticizing of nakedness was done earlier in this century by writers such as Ellis (1901/1942), Fielding (1942), and Fryer (1964). Each of these authors offered extensive surveys of a wide array of aboriginal and tropical cultures that did not require men and women to be as clothed as was required by the modesty traditions of the English-speaking cultures.

The Etiquette of Disregard in the United States
Although nakedness may be more routine in other countries, there are four situations in the United States in which undress is routine. People typically practice an etiquette of disregard in medical situations, nudist camps, art classes, and public rest rooms.

Disregard in medical settings. People in the United States are accustomed to undressing for medical examinations and they expect the medical staff, including orderlies of the opposite sex, to conduct themselves with an etiquette of disregard. The man who would be highly embarrassed to learn that his zipper was open as he discussed his schedule with his appointment secretary can expose a buttock to a female nurse for an injection with a minimum of embarrassment. This is possible in a medical situation because people treat physical exposure in these settings with the etiquette of disregard.

Similarly, a woman who would be mortified if her bikini top dropped before a male acquaintance on the beach would be able to endure a breast or pelvic examination by a male doctor with an etiquette of disregard. A few research studies show that both patients and physicians joke and discuss irrelevant topics in these situations. Emerson's (1970) study is representative. She observed the social interaction of seventy-five gynecological examinations by male doctors and found the participants discussed the procedures as they were happening, and that they minimized the sexuality of the situation by conducting these conversations with a detached and professional style. The physicians gave technical explanations and redefined erotic responses with terms such as "feels ticklish." Others have reported similar findings (Domar, 1985–1986; Henslin and Briggs, 1971; Lief and Fox, 1963). This detached style is learned (Lief and Fox, 1963). Many male physicians describe an initial discomfort with the potential eroticism of pelvic and breast exams (Altucher, 1990). Nevertheless, most appear to adapt to the situation, which they describe as becoming nonerotic for them. One male gynecologist said, "During a pelvic you don't have time to become aroused. You have to do a complicated exam as quickly as possible, so as not to prolong discomfort" (Keyser, 1986, p. 17).

Using this detached style, physicians are typically able to maintain a sufficiently comfortable situation for their women patients. Several surveys of gynecological patients indicate that most of the women experiencing the pelvic examination acclimate to the experience and learn to tolerate it with less embarrassment with both male and female doctors (Haar, Halitsky, and Stricker, 1977; Millstein, Adler, and Irwin, 1984; Vella, 1991).

Disregard in nudist camps. Because of lessons learned through cultural folklore, Americans have learned to think of nudity as sexual and are surprised to learn that most of the participants in nudist camps are not sexually promiscuous (Smith, 1981, p. 75). The data, however, suggest that nudists are usually less sexually experienced than nonnudists and are less likely to engage in sex that might be considered a transgression in nonnudist society than are nonnudists. Story (1987) compared one hundred nudists with a matched sample of one hundred nonnudists and found that social nudists had fewer nontraditional sexual experiences. Based on questionnaire responses, it appears that nonnudists are significantly more likely to have experienced premarital intercourse with someone other than an intended marriage partner, to have experienced homosexual acts, or to have had an extramarital experience. Story's research corroborated the findings of previous studies (Hartman, Fithian, and Johnson, 1971; Weinberg, 1971).

Nudists, or "naturists" as they are often called today, typically minimize sexuality by practicing an etiquette of disregard. It is part of the naturist philosophy to alleviate the idea that "nudity is lewdity" (Goodson, 1991). Nudists learn to focus on their activities, not their bodies, so they do not stare at each other with erotic interest or curiosity (Casler, 1971; Douglas, Rasmussen, and Flanagan, 1977). As a result of not being stared at, people in nudist settings begin to feel more confident and natural, and the potential eroticism in the situation is dispelled. Discussing his first visit to a nudist camp, a conservative religious political advocate, Skipp Porteous, said to an inquirer, "One observes nudity without lusting, in the same way a gardener looks out over the garden without drooling" (Porteous, 1994, p. 14).

Disregard in art classes. Although art students typically anticipate being uncomfortable in the presence of same-sex or opposite-sex nude models in figure drawing classes, reports of art educators indicate that students overcome their embarrassment quickly with what can be called an etiquette of disregard (Champa, 1994; Jesser and Donovan, 1969; Manzella, 1973).

The artist's ability to ignore the erotic potential of the nude model is translated into a painting that de-emphasizes the erotic. In her recent book on the female nude in paintings, Nead (1992) described the typical instruction book for artists painting nude figures. The student painter is tutored in the art of turning the model's nudity into art rather than pornography by systematically de-emphasizing the erotic. A text by Blake (1980) illustrates this kind of instruction. "Avoid 'glamor' poses," the text warned, "that exaggerate the model's charms like a publicity photograph of a star" (p. 65). Similarly, an instruction book by DeRuth (1976) explained, "Don't paint . . . [nipples] in some sweet pink and avoid detail. This would attract attention, suggesting the kind of cheap erotic quality that is least desirable" (p. 54).

Disregard in public rest rooms. Modesty myths tell Americans that they need to protect against sexual arousal in public rest rooms by segregating male and female facilities. The need for such toileting segregation, however, appears exaggerated in cultural myths.

The evidence suggests that people minimize embarrassment or arousal in gender segregated rest rooms by using an etiquette of disregard. Middlemist and his colleagues (Middlemist, Matter, and Knowles, 1976) reported a study of sixty lavatory users in a three-urinal rest room. The closer the researcher stood to the subject, the longer the delay in micturition, suggesting subjects were uncomfortable with the close proximity of others. Reid and Novak (1975) provided additional evidence for this interpretation in a study reporting that, if at all possible, men put one vacant urinal between themselves and anyone else.

Although women's rest rooms typically have stalls that completely enclose them, there is evidence that women, too, use an etiquette of disregard to ease modesty embarrassment when facilities are not enclosed. Vivona and Gomillion (1972) studied women's reactions to dormitory lavatories that had unenclosed toilet and bathtub facilities and found that the women reported adapting to the situation with what might be called an etiquette of disregard.

Research on the etiquette of disregard in public lavatories has been discouraged by the very etiquette that has been the focus of these studies. Shortly after the Middlemist, Matter, and Knowles (1976) study appeared, Koocher (1977) challenged the ethics of researchers making surreptitious observations of urinating behavior in men's lavatories. In effect, Koocher discouraged this kind of study by arguing that the study of the etiquette of disregard violated the etiquette of disregard.

Disregard with homosexuals. Although homosexuals are not barred from public rest rooms and dressing areas, their presence is not easily noticed by heterosexuals, especially by heterosexuals who practice an etiquette of disregard.

For the most part, heterosexuals have a hard time recognizing homosexuals (Berger et al., 1987). Heterosexuals imagine, incorrectly, that the typical gay man has an effeminate face (Dunkle and Francis, 1990; Addison, 1989), and that the typical lesbian is both masculine and unattractive (Dunkle and Francis, 1990). Such illusions are possible because a very large proportion of people in the United States do not know a self-disclosed gay person (Herek and Capitanio, 1996; Turner, 1992; *U.S. News,* 1993). That homosexuals typically pass unnoticed in public toileting settings, however, is strong evidence that both homosexuals and heterosexuals disregard the sexuality of these settings much as medical personnel, art students, and nudists disregard nakedness and minimize the sexuality of their situations.

The ability of homosexuals to suppress their erotic responses to members of the same sex while undressing and toileting with them is enhanced, no doubt, because people generally have some ability to suppress their immediate erotic response (Laws and Holemen, 1978; Mahoney and Strassberg, 1991; Quinsey and Bergersen, 1976; Quinsey and Carrigan, 1978; Rosen, Shapiro, and Schwartz, 1975; Warwick and Salkovskis, 1990) and to treat potential sex objects with an etiquette of disregard.

Summary

All societies must deal with the fact that sexual impulses cannot be given free expression without infringing on the rights of others. But it is possible for a culture to create a misguided folklore as to how these sexual impulses are contained and managed.

People create such folklore when it is imagined that a modesty ceremony serves to contain and diminish sexual impulses. The culture romanticizes the images of nakedness by hiding nakedness between the sexes, except for images that are designed to be sexually alluring. With routine nakedness uncommon, the romanticized pictures become the cultural images of nakedness, creating the illusion that nakedness and undress result naturally in people being seen as sex objects in privacy settings.

When we review settings in which nakedness or toileting are routinized, however, we discover that people in the United States, like people in other cultures, treat the potential sexual opportunities with an etiquette of disregard. Nakedness and toileting in the presence of others is routinized in medical settings, nudist camps, art classes and public rest rooms. These settings are not particularly erotic because the people in these settings find themselves behaving with an etiquette of disregard. In Japan and other countries in which modesty ceremony does not romanticize nakedness, this etiquette is routine. In the United States, these situations are more unusual and there is more awareness and more surprise when people fall quickly into conforming to this etiquette.

Reason tells us that homosexuals are particularly adept at conforming to an etiquette of disregard. The typical homosexual has spent a lifetime dressing and toileting with people of the same sex and has learned to disregard any potential sexual allure of heterosexuals in this context. Homosexuals in the United States have not romanticized nakedness with heterosexuals in the way that heterosexual men have romanticized female nakedness. Homosexuals understand that heterosexuals are typically not receptive to homosexual advances, and they are motivated to conform to this etiquette even though heterosexual disapproval is the only sanction in civilian life against minor violations in privacy settings.

If homosexuals are able to manage their sexual response in the presence of undressed heterosexuals so as to minimize embarrassment for everyone, then it is unnecessary to ban them from the military to protect the modesty concerns of heterosexuals.

Note

1. General Colin Powell said, "To introduce a group of individuals who—proud, brave, loyal, good Americans—but who favor a homosexual lifestyle, and put them in with heterosexuals who would prefer not to have somebody of the same sex find them sexually attractive, put them in close proximity, ask them to share the most private of their facilities together, the bedroom, the barracks, latrines, the showers, I think that's a very difficult problem to give the military." This quotation is taken from the ABC News *Nightline* transcript for Show #2867 (p. 1), which aired May 19, 1992. According to the transcript, Powell actually made this statement on February 5, 1992.

References

Abele, A. (1986). Functions of gaze in social interaction: Communication and monitoring. *Journal of Nonverbal Behavior, 10,* 83–101.

Addison, W. E. (1989). Beardedness as a factor in perceived masculinity. *Perceptual & Motor Skills, 68,* 921–922.

Altucher, B. (1990). Women's health, men's work. *Health, 22,* 60–64.

Antoun, R. T. (1968). On the modesty of women in Arab Muslim villages: A study of the accommodation of traditions. *American Anthropologist, 70,* 671–697.

Argyle, M. (1975). *Bodily communication.* London: Methuen.

Asayama, S. (1976). Sexual behavior in Japanese students: Comparisons for 1974, 1960, and 1952. *Archives of Sexual Behavior, 5,* 371–390.

Asendorpf, J. B. (1990). The expression of shyness and embarrassment. In W. R. Crozier (Ed.), *Shyness and embarrassment: Perspectives from social psychology* (pp. 87–118). Cambridge: Cambridge University Press.

Berger, G., Hank, L., Rauzi, T., & Simkins, L. (1987). Detection of sexual orientation by heterosexuals and homosexuals. *Journal of Homosexuality, 13,* 83–100.

Blake, W. (1980). *Figures in oil.* New York: Watson Guptill.

Casler, L. (1971). Nudist camps. *Medical Aspects of Human Sexuality, 5,* 92–99.

Champa, P. (1994). Painting from the model. *American Artist, 58,* 70–76.

Crawley, E. (1965). Nudity and dress. In M. E. Roach & J. B. Eicher (Eds.), *Dress, adornment, and the social order* (pp. 46–49). New York: Wiley.

Cupach, W. R., Metts, S., & Hazleton, V. (1986). Coping with embarrassing predicaments: Remedial strategies and their perceived utility. *Language and Social Psychology, 6,* 181–200.

Dekker, J., & Everaerd, W. (1988). Attentional effects on sexual arousal. *Psychophysiology, 25* (1), 45–54.

————. (1989). Psychological determinants of sexual arousal: A review. *Behaviour Research & Therapy, 27* (4), 353–364.

DeRuth, J. (1976). *Painting the nude.* New York: Watson Guptill.

Domar, A. D. (1985–1986). Psychological aspects of the pelvic exam: Individual needs and physician involvement. *Women & Health, 10,* 75–90.

Douglas, J. D., Rasmussen, P. K., & Flanagan, C. A. (1977). *The nude beach.* Beverly Hills, CA: Sage.

Downs, J. F. (1990). Nudity in Japanese visual media: A cross-cultural observation. *Archives of Sexual Behavior, 19,* 583–594.

Dunkle, J. H., & Francis, P. L. (1990). The role of facial masculinity/femininity in the attribution of homosexuality. *Sex Roles, 23,* 157–167.

Edelmann, R. J. (1990). Embarrassment and blushing: A component process model, some initial descriptive and cross-cultural data. In W. R. Crozier (Ed.), *Shyness and embarrassment: Perspectives from social psychology* (pp. 205–229). Cambridge: Cambridge University Press.

————. (1994). Embarrassment and blushing: Factors influencing face-saving strategies (pp. 231–267). In S. Ting-Toomey (Ed.), *The challenge of facework: Cross-cultural and interpersonal issues.* Albany, NY: State University of New York Press.

Edelmann, R. J., Asendorpf, J., Contarello, A., Georgas, J., Villanueva, C., & Zammuner, V. (1987). Self-reported expression of embarrassment in five European cultures. *Social Science Information, 26,* 869–883.

Edelmann, R. J., & Hampson, S. E. (1981). Embarrassment in dyadic interaction. *Social Behavior and Personality, 9,* 171–177.

Edelmann, R. J., & Iwawaki, S. (1987). Self-reported expression and consequences of embarrassment in the United Kingdom and Japan. *Psychologia: An International Journal of Psychology in the Orient, 30,* 205–216.

Ellis, H. (1942). The evolution of modesty. *Studies in the psychology of sex,* Vol. 1. New York: Random House. (Original work published in 1901).

Emerson, J. P. (1970). Behavior in private places: Sustaining definitions of reality in gynecological examinations. *Recent Sociology, 2,* 74–97.

Exline, R. V., Gray, D., & Schuette, D. (1965). Visual behavior in a dyad as affected by interview content and sex of respondent. *Journal of Personality and Social Psychology, 3,* 201–209.

Fielding, W. J. (1942). *Strange customs of courtship and marriage.* New York: New Home Library.

Flaming, D., & Morse, J. M. (1991). Minimizing embarrassment: Boys' experiences of pubertal changes. *Issues in Comprehensive Pediatric Nursing, 14,* 211–230.

Fryer, P. (1964). *Mrs. Grundy: Studies of English prudery.* New York: London House & Maxwell.

Goldman, R. J., & Goldman, J. D. (1981). Children's perceptions of clothes and nakedness: A cross-national study. *Genetic Psychology Monographs, 104* (9), 163–185.

Goodson, A. (1991). *Therapy nudity and joy: The therapeutic use of nudity through the ages from ancient ritual to modern psychology.* Los Angeles: Elysium Growth Press.

Haar, E., Halitsky, V., & Stricker, G. (1977). Patients' attitudes toward gynecologic examinations and to gynecologists. *Medical Care, 15,* 782–790.

Hartman, W., Fithian, M., & Johnson, D. (1971). *Nudist society.* New York: Avon.

Henslin, J., & Briggs, M. (1971). Dramaturgical desexualization: The sociology of the vaginal examination. In J. Henslin (Ed.), *Studies in the sociology of sex* (pp. 243–277). New York: Appleton-Century Crofts.

Herek, G. M., & Capitanio, J. P. (1996). "Some of my best friends": Intergroup contact, concealable stigma, and heterosexuals' attitudes toward gay men and lesbians. *Personality and Social Psychology Bulletin, 22,* 412–424.

Imahori, T. T., & Cupach, W. R. (1994). A cross-cultural comparison of the interpretation and management of face: U.S. American and Japanese responses to embarrassing predicaments. *International Journal of Intercultural Relations, 18,* 193–219.

Jesser, C., & Donovan, L. (1969). Nudity in the art training process: An essay with reference to a pilot study. *Sociological Quarterly, 10,* 355–371.

Katz, L. (1989, May). Nudity at home. *Parent's Magazine, 64,* 208.

Keyser, H. (1986). *Women under the knife.* New York: Warner Books.

Kinsey, A., Pomeroy, W. B., & Martin, C. E. (1948). *Sexual behavior in the human male.* Philadelphia: W.B. Saunders.

Kitazawa, K., & Beerman, M. S. (1993). Sexuality issues in Japan: A view from the front on HIV/AIDS and sexuality education. *Siecus Report, 22,* 7–12.

Kleinke, C. L. (1986). Gaze and eye contact: A research review. *Psychological Bulletin, 100,* 78–100.

Koocher, G. P. (1977). Bathroom behavior and human dignity. *Journal of Personality and Social Psychology, 35,* 120–121.

Koukounas, E., & Over, E. (1993). Habituation and dishabituation of male sexual arousal. *Behavior Research Therapy, 31,* 575–585.

Laws, D. R., & Holemen, M. L. (1978). Sexual response faking by pedophiles. *Criminal Justice and Behavior, 5,* 343–356.

Leonard, J. (1992, May). Is modesty the best policy? *Parents Magazine, 67,* 132.

Lewis, R. J., & Janda, L. H. (1988). The relationship between adult sexual adjustment and childhood experiences regarding exposure to nudity, sleeping in the parental bed, and parental attitudes toward sexuality. *Archives of Sexual Behavior, 17,* 349–362.

Lief, H. I., & Fox, R. (1963). Training for detached concern. In H. I. Lief, V. F. Lief, & N. R. Lief (Eds.), *The psychological bases of medical practice* (pp. 12–35). New York: Harper & Row.

Maccoby, E., & Jacklin, C. (1974). *The psychology of sex differences.* Palo Alto, CA: Stanford University Press.

Mahoney, J. M., & Strassberg, D. S. (1991). Voluntary control of male sexual arousal. *Archives of Sexual Behavior, 20,* 1–16.

Malchow, C. W. (1923). *The sexual life.* St. Louis: C.V. Mosby.

Manzella, D. B. (1973). Nude in the classroom. *American Journal of Art Therapy, 12,* 165–182.

Matsubara, L. (1994). Comics for adults: Many Japanese are scandalized by erotic love scenes drawn by unconventional cartoonist. *Far Eastern Economic Review, 157,* 70.

Meuwissen, I., & Over, R. (1990). Habituation and dishabituation of female sexual arousal. *Behavior Research Therapy, 28,* 217–226.

Middlemist, R. D., Matter, C. F., & Knowles, E. (1976). Personal space invasions in the lavatory: Suggestive evidence for arousal. *Journal of Personality and Social Psychology, 33,* 541–546.

Millstein, S. G., Adler, N. E., & Irwin, C. E. (1984). Sources of anxiety about pelvic examinations among adolescent females. *Journal of Adolescent Health Care, 5,* 106–111.

Moore, B. (1984). *Privacy: Studies in social and cultural history.* Armonk, NY: M.E. Sharpe.

Muir, F. (1983). *An irreverent and almost complete social history of the bathroom.* New York: Stein & Day.

Nead, L. (1992). *The female nude: Art, obscenity and sexuality.* New York: Routledge.

O'Donohue, W. T., & Geer, J. H. (1985). The habituation of sexual arousal. *Archives of Sexual Behavior, 14,* 233–246.

Parke, R. D., & Sawin, D. B. (1979). Children's privacy in the home: Developmental, ecological and child-rearing determinants. *Environment and Behavior, 111,* 87–104.

Perin, N. (1969). *Dr. Bowdler's legacy: A history of expurgated books in England & America.* New York: Atheneum.

Porteous, S. (1994). Baring the threat. *Free Inquiry, 14,* 16.

Quinsey, V. L. & Bergersen, S. G. (1976). Instructional control of penile circumference in assessments of sexual preference. *Behavior Therapy, 7,* 489–493.

Quinsey, V. L., & Carrigan, W. F. (1978). Penile responses to visual stimuli: Instructional control with and without auditory sexual fantasy correlates. *Criminal Justice and Behavior, 5,* 333–341.

Reid, E., & Novak, P. (1975). Personal space: An unobtrusive measures study. *Bulletin of the Psychonomic Society, 5,* 265–266.

Rosen, R. C., Shapiro, D., & Schwartz, G. E. (1975). Voluntary control of penile tumescence. *Psychosomatic Medicine, 37,* 479–483.

Rosenfield, A., Siegel-Gorelick, B., Haavik, D., Duryea, M., Wenegrat, A., Martin, J., & Bailey, R. (1984). Parental perceptions of children's modesty: A cross-sectional survey of ages two to ten years. *Psychiatry, 47,* 351–365.

Rugoff, M. (1971). *Prudery and passion: Sexuality in Victorian America.* New York: G.P. Putnam's Sons.

Scheman, J. D., & Lockard, J. S. (1979). Development of gaze aversion in children. *Child Development, 50,* 594–596.

Schulz, R., & Barefoot, J. (1974). Nonverbal responses and affiliative conflict theory. *British Journal of Social and Clinical Psychology, 13,* 1–7.

Seabrook, J., & Burchill, J. (1994, August 12). Keep your shirt on. *New Statesman & Society, 7,* 22–25.

Sears, R., Maccoby, E., & Levin, H. (1957). *Patterns of child rearing.* Evanston, IL: Row, Peterson.

Segal, J., & Segal, Z. (1990, May). Standards on nudity, *Parents Magazine, 65,* 211.

Shawver, L. (1995). *And the flag was still there: Straight people, gay people, and sexuality in the U.S. military.* Binghamton, NY: Haworth Press.

Shelley, S. (1981). Adolescent attitudes as related to perception of parents and sex education. *The Journal of Sex Research, 17,* 350–367.

Shimazaki, T., & Kaji, Y. (1993). A closer look at sexuality education and Japanese youth. *Siecus Report, 22,* 12–16.

Smith, D. C. (1981). *The naked child: The long-range effects of family and social nudity.* Palo Alto, CA: R & E Research Associates.

Smith, H. (1979). A modest test of cross-cultural differences in sexual modesty, embarrassment and self-disclosure. *Qualitative Sociology, 3,* 223–241.

Steffan v. Cheney, Civ. Act. No. 88-3669-Og, at 27–28 (D.D.C. Dec. 9, 1991).

Stephens, W. N. (1971). *A cross-cultural study of modesty and obscenity and pornography* (Vol. 3). Washington, DC: U.S. Government Printing Office.

Story, M. D. (1987). A comparison of social nudists and non-nudists on experience with various sexual outlets. *The Journal of Sex Research, 23,* 197–211.

Taylor, S., & Baker, J. F. (1992, September 14). Madonna's "Sex" too hot for Japanese publisher. *Publishers Weekly,* 12.

Toufexis, A. (1993, May 24). Sex has many accents. *Time,* 66.

Turner, B. (1992, September 14). Gays under fire. *Newsweek,* 35–40.

U.S. News & World Report press release. (1993, June 26).

Vella, P. V. (1991, November). A survey of women undergoing a pelvic examination. *Australian and New Zealand Journal of Obstetrics and Gynecology, 31,* 355–357.

Vivona, C., & Gomillion, M. (1972). Situation morality of bathroom nudity. *The Journal of Sex Research, 3,* 123–135.

Walters, R. G. (1974). *Primers for prudery: Sexual advice to Victorian America,* Englewood Cliffs, NJ: Prentice-Hall.

Warwick, H. M., & Salkovskis, P. M. (1990). Unwanted erections in obsessive-compulsive disorder. *British Journal of Psychiatry, 157,* 919–921.

Weinberg, M. (1964). Sexual modesty, social meanings, and the nudist camp. *Social Problems, 12,* 311–318.

———. (1971, August). Nudists. *Sexual Behavior,* 51–55. Cited in Story, M. D. (1987). A comparison of social nudists and non-nudists on experience with various sexual outlets. *The Journal of Sex Research, 23,* 197–211.

Whiting, J. W. M., & Child, I. L. (1953). *Child training and personality: A cross-cultural study.* New Haven, CT: Yale University Press.

Zillman, D., & Bryant, J. (1988). Pornography's impact on sexual satisfaction. *Journal of Applied Social Psychology, 18,* 438–453.

PART FOUR

Implementation

Issues of Confidentiality: Therapists, Chaplains, and Health Care Providers

Jeffrey E. Barnett and Timothy B. Jeffrey

Fundamental to the delivery of health care, counseling, and psychotherapy is a relationship of trust. Patients (hereafter referred to as consumers) share personal and private information with providers in an effort to receive desired and necessary assistance. Withholding or intentionally distorting important information undermines the therapeutic process. It is widely understood that consumers of health care services would be reluctant to speak freely if disclosures of personal and private information by providers were imminent. During therapy, consumers often share fantasies, ideas, beliefs, and attitudes that would be embarrassing or compromising, or would even lead to serious repercussions, such as the loss of relationship or employment, if they were made known to others. Often it is through the sharing of such personal information that treatment occurs. To the degree that respect for the confidentiality of consumer communications is compromised, so, too, is the treatment process.

For example, a psychotherapy consumer learning to work through anger toward a supervisor might reveal fantasies of aggression and revenge. The disclosure of such fantasies to the employer would jeopardize not only the treatment process but the individual's employment status as well. Military service members wishing to explore issues of sexuality and sexual orientation, especially if they were to reveal the presence of a sexual orientation not held by the majority, might face a similar dilemma in treatment. In such situations, trust and confidentiality are crucial to effective treatment. As demonstrated in numerous cases in the media, disclosure of such information by U.S. military personnel typically produces dire consequences, essentially ending their careers as soldiers, sailors, marines, and airmen. It should be noted that such is not the case for military personnel in many foreign countries (see chapter 6).

Entering into counseling or psychotherapy is frequently stigmatized and viewed spuriously. Were an individual's involvement in such processes likely to become public knowledge, many persons in need of such services would choose to forgo them. As Reisner (1985) stated, "the promise of confidentiality, both as to the fact that individuals are in psychotherapy or counseling as well as to the contents of their disclosures, is thus important" (p. 203).

Historical Underpinnings

The obligation of health care providers to maintain confidentiality has great historical precedence. As stated in the Hippocratic Oath, "Whatever, in connection with my professional practice, or not in connection with it, I see or hear, in the life of men, which ought not to be spoken of abroad, I will not divulge, as reckoning that all such should be kept secret" (in Gutheil and Appelbaum, 1982, p. 5).

An individual's right to privacy is well rooted in American legal tradition. In 1890 Warren and Brandeis published "The right to privacy," a journal article that became the basis for many judicial decisions and legislation (Meyer, Landis, and Hays, 1988). As Shah (1969) described it, "the concept of privacy recognized the freedom of the individual to pick and choose for himself the time, circumstances, and particularly the extent to which he wishes to share with or withhold from others his attitudes, beliefs, behavior, and opinions" (p. 57).

Further precedence for the right to privacy is found in the Fifth Amendment of the Constitution of the United States. Moreover, a clergy-parishioner privilege is mentioned in the First Amendment. In 1965, the United States Supreme Court in *Griswold v. Connecticut* "recognized a concrete application of the legal right to privacy was guaranteed by the Constitution" (in Stromberg et al., 1988, pp. 374–375). Numerous subsequent rulings have protected individuals from both government intrusion (see *Roe v. Wade,* 1973; *Barefoot v. Estelle,* 1983) and from intrusions by others (see *Doe v. Roe,* 1977; *Vassiliades v. Garfinckels,* 1985).

Chaplains and health care professional associations have long maintained confidentiality requirements in their codified standards for professional practice. Organizations such as the American Medical Association, American Psychiatric Association, American Psychological Association, National Association of Social Workers, American Nurses Association, and the American Association of Pastoral Counselors emphasize the sanctity of confidentiality within professional relationships. By so doing they reify the fundamental importance of confidentiality to the treatment process.

The Code of Ethics of the American Psychological Association (APA, 1992) stresses the need to "accord appropriate respect to the fundamental rights, dignity, and worth of all people." Moreover, psychologists are instructed to "respect the rights of individuals to privacy, confidentiality, self-determination, and autonomy, mindful that legal and other obligations may lead to inconsistency and conflict with the exercise of these rights" (APA, 1992, Principle D: Respect for People's Rights and Dignity, p. 1599). The significance of "legal and other obligations" that may lead to "inconsistency and conflict with the exercise of these rights" will become apparent as conflicts between ethical obligations and military regulations are reviewed later in this chapter. Similarly, the code of ethics of the American Medical Association (1973) allows breaches of confidentiality only when compelled by law or when necessary to protect the welfare of the consumer or community. Psychologists as well are specifically instructed to "take reasonable precautions to respect the confidentiality rights of those with whom they work or consult, recognizing that confidentiality may be established by law, institutional rules, or professional or scientific relationships" (APA, 1992, Standard 5.02: Maintaining Confidentiality, p. 1606).

In addition to being ethically bound to maintain standards of confidentiality, health care providers are mandated by state statutes to maintain such standards. Review of licensure and certification statutes suggest that such regulation inherently obligates "practitioners to comply with professional ethical standards" (Reisner, 1985, p. 265). Strong precedents exist for this position (e.g., *Morra v. State Board of Examiners of Psychologists,* 1973). This is most relevant to the military because uniformed health care providers must be licensed or certified in their profession to provide treatment services (DoD, 1985).

The obligation not to disclose information gathered within the confines of the professional relationship has been established and refined in both statutes and case law. As Cohen and Mariano (1982) stated, "States

... have enacted statutes that have extended the right of the individual to disclose certain information." This right "has been extended to those persons with whom the individual shares a special relationship, e.g., attorney, physician, psychologist, priest" (p. 260).

Privilege

Privilege is a concept related to confidentiality. Whereas confidentiality refers to the protection afforded information shared in certain special relationships, privilege serves as a protection from disclosure of such information in a court of law. All states have statutes governing privilege, which define the rights of an individual in legal proceedings. New York, for example, provides in its laws of evidence that "unless the patient waives the privilege, a person authorized to practice medicine or dentistry, or a registered professional or licensed practicing nurse, shall not be allowed to disclose any information which he acquired in attending a patient in a professional capacity, and which was necessary to enable him to act in that capacity" (New York Evidence Law, 1962). Such statutes and the relevant case law have been revised to include psychologists, psychotherapists, and clergy. These statutes identify the consumer as the holder of privilege and thus the one who maintains control over the release of any information shared within the confines of these special relationships.

Privilege is a special case of confidentiality that refers to the lawful basis for a refusal to release information from a therapeutic relationship in legal proceedings. Several exceptions to therapist-patient privilege exist. For example, a court may order a litigant to undergo a psychological evaluation. All parties understand that the results of such an evaluation will be made known to the court. A second example is found in the patient-litigant exception to the testimonial privilege laws (Weiner and Wettstein, 1993, p. 213). Under this exception, when a litigant enters one's mental status or functioning as an element of a claim in a legal proceeding, privilege is waived so the court may evaluate all relevant facts. In this situation, all legal testimony, by definition, becomes a matter of public record.

Breach of Confidentiality

As previously discussed, case law, statutes, and professional ethical codes all establish the right to privacy and the obligation to safeguard information shared with clergy, physicians, psychologists, and therapists. As with privilege, however, exceptions to the rule of confidentiality exist. In general, professional ethics codes defer to existing laws, regulations, and organizational rules when examining the issue of confidentiality.

Many consumers assume that all information shared within the context of the professional relationship is confidential. Many health care providers may assume this as well. They may advise consumers of this, only to be surprised when exceptions and limits of confidentiality are discovered. In actual practice, numerous exceptions to absolute confidentiality exist. It is vitally important that all individuals be aware of these limits when they enter treatment so as to avoid a plethora of potential conflicts and dilemmas. The Manual for Courts-Martial (DoD, 1994), for example, makes it clear that there are limits of confidentiality for military health care providers. It also provides that penitent-clergy communications are privileged when made either as a religious act or as a matter of conscience (Trower, 1992). In this respect, clergy may have greater confidentiality of communications than military health care providers. The rules and policies regarding confidentiality vary among churches, and in that regard military chaplains have a certain obligation to be clear on any limits of confidentiality present in a particular situation (see, e.g., Jacob, 1982; Creswell, 1981).

The courts have long attempted to balance the preservation of the individual's right to confidentiality with protection of the public. In a defining case, the California Supreme Court ruled that psychotherapists and health care providers are obligated to keep communications confidential unless disclosure is made specifically to avoid harm to others. The ruling stated, "protective privilege ends where the public peril begins" (*Tarasoff v. Regents of University of California,* 1976). Further rulings refined this exception to cases in which a specific threat to do harm to an identifiable victim is made in the context of the psychotherapy or treatment relationship (see *Thompson v. County of Alameda,* 1980; *Lipari v. Sears, Roebuck & Co.,* 1980; and *McIntosh v. Milano,* 1979). Such requirements have also been codified in statutes in many states (e.g., in Maryland, Confidentiality of Medical Records, 1991).

Other exceptions to confidentiality that are regulated by statute include the requirement to report suspected child abuse or neglect and, in some jurisdictions, spouse and elder abuse. Moreover, several statutes provide for disclosures without the authorization of the consumer. These include sharing information with other health care providers for purposes, such as coordinating health care services and pursuing involuntary commitment proceedings (e.g., Confidentiality of Medical Records, 1991).

The above discussion notwithstanding, the most common exception to confidentiality occurs with consumer consent. For a consent to be valid, Stromberg et al. (1988) recommended that consent be documented in writing, and that the consumer understand the specifics and scope of the information to be disclosed, with whom it would be shared, and the general purpose(s) for which it would be provided (p. 391). Many jurisdictions require that consent be documented in writing so as to verify understanding of the scope and purpose of such consent.

It is incorrect to assume that consumers accurately understand confidentiality and its limits. One study (Miller and Thelen, 1986) found that 69 percent of survey respondents believed that all material discussed in psychotherapy was confidential. Moreover, 97 percent expressed the desire to be informed of limits of confidentiality prior to onset of treatment. Failure to make such a disclosure was particularly troubling in the case of Corporal Kevin Blaesing, a young, active-duty Marine who questioned his sexual orientation during a therapy session with a military psychologist (Raddatz, 1994). Much to Blaesing's dismay, the psychologist breached confidentiality and told the service member's commander. Military regulations do *not* require such disclosure for referred consumers. The result was predictable: His fitness for duty was deemed poor and he was recommended for discharge by a board of Marine Corps officers. This decision was overturned on appeal. Due to the notoriety of the case, however, it is questionable whether he will be able to complete his military career successfully.

Blaesing was never advised of any limits of confidentiality, and, in fact, assumed that everything said in the doctor-patient relationship was held in the strictest confidence. Concern regarding the need for military health care providers to advise consumers about the limits of confidentiality has been emphasized for several years (Jeffrey, 1989). The failure of the health care provider in this case, one hopes, was an isolated, atypical situation and does not reflect normative behavior for military psychologists. It is interesting to note that even the client's commanding officer, Lieutenant Colonel Martin J. Martinson, thought that all conversations with the military health care providers were strictly confidential. As confirmed by an unnamed Department of Defense (DoD) official, such is not the case. According to that official, there is no promise of confidentiality in the military for information shared with DoD health care professionals (Raddatz, 1994).

Clear guidance is provided for addressing the release of confidential information. For example, the General Guidelines for Providers of Psychological Services (APA, 1987) states: "Psychologists do not release confidential information, except with the written consent of the user involved, or of his legal representative, guardian, or other holder of the privilege on behalf of the user, and only after the user has been assisted to understand the implications of the release" (p. 21). Further guidance is provided in the Ethical Principles of Psychologists and Code of Conduct (APA, 1992), which emphasizes the need to inform consumers *and organizations* "with whom they establish a scientific or professional relationship . . . of (1) the relevant limitations on confidentiality, including limitations where applicable . . . and (2) the foreseeable uses of the information generated through their services" (p. 1606). Moreover, "In order to minimize intrusions on privacy, psychologists include in written and oral reports, consultations, and the like, only information germane to the purpose for which the communication is made" (p. 1606). It is important to note that a frequent error occurs when providers obtain consent to release confidential information, but then disclose more information than was authorized by the consumer. The following example highlights this point.

A military aviator was grounded while receiving treatment for injuries suffered during an incident in which the aviator's copilot was killed. The commander ordered a psychological evaluation to determine if therapy was warranted and to obtain an evaluation of fitness for return to duty. No treatment was needed, and the evaluation was successfully completed, but in the report to the commander, the psychologist provided information the aviator shared in passing concerning the commander's leadership style. These statements were less than complimentary and not well received by the commander. As a result, the aviator was not returned to flight status and was relegated to a support position.

As an experienced officer, the aviator would never have made such statements outside the "protected environment" of the psychotherapeutic relationship. A release had been signed that listed the limits of confidentiality. The aviator, however, rightly understood that only information relevant to the current evaluation and to the ability to return to active flight status would be released. During the evaluation, the psychologist encouraged discussion of the commander's leadership style. The aviator rightly assumed that such information was incidental to the process and would be kept in confidence. Additionally, if the commander expressed an expectation that such information would be released, it would be the psychologist's duty to advise the commander of the limits of confidentiality. This example highlights the need for both the consumer and organization to be fully informed in advance of the limits of confidentiality, so that each will have appropriate expectations of the process and so that the consumer will fully grasp the implications of sharing information before doing so. Specific guidance for this is seen in the American Psychological Association ethics code, which states, "Psychologists discuss confidential information obtained in clinical or consulting relationships, or evaluative data concerning patients, individual or organizational clients . . . only for appropriate scientific or professional purposes and only with persons clearly concerned with such matters" (APA, 1992, p. 1606).

In this case, unit policy might render it appropriate for the psychologist to inform the base finance office if the aviator were not returned to flight status and no longer to receive flight pay. It would not, however, be appropriate to discuss with finance personnel specifics of the evaluation or reasons for such a decision. It would not be relevant to the change in pay status to share clinical or otherwise personal and private information about the aviator/patient. Numerous authors (e.g., Pope, 1990) stress the importance of "divulging confidential information only to the extent required by law" (p. 40).

This example demonstrates the need for both the consumer and the organization to understand all relevant limits to confidentiality, whether they are set by ethics, law, regulation, or institutional policy. The explanation of these limits is of little, if any, value if received *after* sensitive material is shared, and should be reviewed at the beginning of any consultation or therapeutic contact. To this end, Pope further stresses the need for practitioners "to clarify what their legal (according to legislation and case law applicable in their state) and ethical obligations are to patients in this regard. Not only the legal and ethical, but also the clinical implications of withholding this information from patients need to be considered" (Pope, 1990, p. 41). The obligation to discuss the limits to confidentiality comprehensively before disclosures are made is seen in standards of professional ethics such as those published by psychologists (APA, 1992), which state: "Unless it is not feasible or is contraindicated, the discussion of confidentiality occurs at the outset of the relationship and thereafter as new circumstances may warrant" (p. 1606).

Although professional ethics codes do provide general guidance on issues such as confidentiality for health care providers, they are not designed to address or anticipate all possible cases, scenarios, dilemmas, or conflicts that might arise. According to Pope and Vasquez (1991),

> Ethics codes, standards, or rules can never legitimately serve as a substitute for a thoughtful, creative, and conscientious approach to our work. They can never relieve us of the responsibility to struggle with competing demands, multiple perspectives, evolving situations and the prospect of uncertain consequences . . . [they] serve best to awaken us to the potential pitfalls, but also to opportunities to guide and inform our attempts to help without hurting. They cannot do our work for us. (Pp. 49–50)

Professional ethics codes and standards provide necessary, but often not sufficient, guidance for military health care providers to make appropriate decisions. Such codes (e.g., APA, 1992, 1987) guard consumer confidentiality when such behavior is consistent with relevant laws, rules, and regulations. Weiner and Wettstein (1993) stressed to health care providers that "adhering to the clinical standard of care will not always immunize the clinician against liability. Various administrative policies, legal regulations, and case law, or statutes may demand even more of the clinician. Following these policies is critical" (p. 183).

Historically, ethics codes were viewed as above or superior to law, but in recent years there has been an ascendancy of law over ethics. One example of this is seen in the concept of limits of confidentiality. By deferring to laws and regulations, ethics codes undermine the basis of the concept of confidentiality. Even clergy, who in the past justified their practices by deferring to a higher order, tailor their practices to conform to legal precedent (Denham and Denham, 1986).

The Military Setting

For the past fifty years, the military has been populated by young, physically healthy, relatively well-educated, action-oriented individuals who were often geographically separated from members of their families. Most agree that service in the military is stressful, especially for first-term enlistees, those assigned to geographically isolated hardship areas, and those who, because of the nature of their duties or unit's mission, are subjected to greater demands than typically would be found in most forms of civilian employment. The need for, and availability of, counseling in such a setting is obvious. The role of the military chaplain is especially important in such an environment, in part because of the high percentage of young men and women who were exposed to clergy as part of their religious and educational development and, as a result, expect to be able to rely upon them for counseling and other services. In the military the directive "Go see the chaplain" has been used as a pejorative injunction, a joke, a helpful suggestion, a put-off, and for multiple other purposes (Trower, 1992). The only reason one almost never is told to "Go see the psychologist, psychiatrist, or counselor" is that these mental health practitioners often do not have the same level of unit involvement as the chaplain. Many units have an assigned chaplain, but such is not the case for mental health practitioners. Many unit commanders are probably very happy to keep it that way.

Gutheil and Appelbaum (1982) discuss the issue of agency, which they describe as "for whom one is working: who as employer has hired the therapist. The agency is thus the operational basis for the therapeutic alliance" (p. 14). Inherent in this is that "the clinician's agency is usually split in varying proportions between the individual and the institution. Examples of a split therapeutic agency might occur in military, court, school, or industrial mental health work" (p. 15).

For providers employed by an agency there is, in fact, allegiance and duty owed both to the consumer and to the organization. How this allegiance is split and how one's duty to each is demonstrated are subjects of discussion and debate. Relevant regulations provide guidance, but ambiguity exists for health care providers to resolve. Although military clergy and health care providers may defend the need for primary obligation to the military organization, ethical codes stress the need to protect individual rights and work to serve the best interests of each consumer. Conversely, whereas a health care provider may retain a primary obligation to the consumer, relevant organizational rules and regulations may not be ignored. The balancing of these competing demands has been described as striving to serve two masters (Jeffrey, Rankin, and Jeffrey, 1992). Others have labeled it double agentry (Hastings Center, 1978; Simon, 1992). Further, this problem is not restricted solely to the military (Sharkin, 1995).

Legitimate needs of the military often contradict professional ethical standards. All health care providers employed by the DoD, for example, are bound by regulations that govern confidentiality and its exceptions in serving military personnel (e.g., DoD, 1975). Specific agency regulations restrict confidentiality in the various services (e.g., Department of the Army (DA), 1987). As stated by Jeffrey et al. (1992, p. 91), "Although federal statutes, DoD directives, and service regulations are written so as to respect the privacy of individuals, they also mandate access to confidential material by federal employees with a 'need for the record in the performance of their duties'" (DA, 1982b, 1985; DoD, 1975). These regulations do not require that advance notice or consent be provided by the service member. Those serving in the military do not receive the same protection as civilians in doctor-patient relationships, and they may not even be informed of such limitations (e.g., Raddatz, 1994). They may learn of such only after information has been released, and as a result of the consequences of such release.

Pope (1990) provides an illustrative anecdote described originally by Dyer (1988) that holds special relevance to confidentiality issues for gays and lesbians in the military.

> At a lecture I gave on confidentiality, a physician in the audience related the following story from his experience as a medical officer during the Korean War: A soldier reported to the infirmary with feelings of depression and an inability to sleep, eat, or concentrate on his work. He said he was upset because his homosexual lover had rejected him. He expressed thoughts of suicide. The physician referred the soldier to the nearby psychiatric unit. There the admitting officer met the soldier, not with the usual offer to help, but with the announcement that this would mean an end to the soldier's military career. Shortly after that the soldier shot and killed himself. (P. 68)

This example illustrates the need to inform the consumer in advance of the limits of confidentiality, the nature and potential uses of the information to be released, and the grave consequences that may result when such ethical obligations are not met adequately. One way to introduce this concept to a consumer is with a specific written agreement, such as the one provided in the appendix.

As federal agencies, the various branches of the military have at their disposal regulations allowing organizational representatives access to otherwise confidential material because of a need-to-know clause (e.g., DA, 1981, 1982). The privacy act that provides this authority states: "No agency shall disclose any record . . . unless disclosure of the record would be (1) to those officers and employees of the agency which maintains the record who have a need for the record in the performance of their duties" (Privacy Act, 1974). These regulations do not defer to professional ethics codes. Moreover, they substantially undermine the consumer as holder of the privilege. Although civilian consumers typically have the many rights previously discussed, in the military the organization and its representatives become the de facto holders of the privilege. This is so not only for active duty military personnel, but also for their family members seen in military medical department facilities. Commanders with a need to know are granted free access to information shared in therapy or health care relationships. For this reason, it is important that military health care providers carefully attend to the information that they place in health care records. President Clinton's "Don't Ask, Don't Tell" policy does *not* prohibit officials within DoD from access to military health care records. Need to know is not defined in DoD directives or military regulations. It is a broadly granted power and justified because of national defense. DoD officials are entrusted to exercise their powers appropriately regarding this policy. It was the experience of the authors of this chapter that almost all officials within the DoD did not abuse this power. Nevertheless, it is important to remember that the power exists and that no notice to the consumer or written consent is required before gaining access to information shared with DoD health care providers.

These issues have a clear and profound impact for homosexual service members. Although all service members are legally afforded the same health care benefits, the limits to confidentiality may remove military health care providers as a viable resource for gay and lesbian service members. At the least it creates a special requirement to educate service members and military health care providers about limits of confidentiality and regulations that constrain the freedoms of military personnel (e.g., Howe, 1989). Because military health care providers retain some autonomy and decision-making ability, their education is important. Health care providers who are sympathetic, or at least sensitized, to the issues and needs of gays and lesbians will likely endeavor to give adequate attention to service members' needs when complying with military directives. For example, providers will discuss the limits of confidentiality so as to assist consumers in making informed decisions about disclosure of information.

In the past, homosexuality was viewed as a form of psychopathology and sexual deviancy (APA, 1952). Only in the past twenty years have medical, psychiatric, psychological and some religious communities accepted it as a normal lifestyle (APA, 1994). The Diagnostic and Statistical Manual of Mental Disorders (American Psychiatric Association, 1994) descriptions highlight how mental health professionals' views of homosexuality have matured in recent years. Unfortunately, the military has lagged in its understanding of this issue. As discussed in chapter 6, the American military deviates from that of many other nations and continues to operate on questionable assumptions about the behavior of persons with a homosexual orientation. As documented in earlier chapters, much empirical data support a position of accepting gays and lesbians in all aspects of military life, but for many, the military remains a hostile environment. President Clinton's well-publicized difficulties in his attempt to remove the ban on gays and lesbians, and the resultant "Don't Ask, Don't Tell" compromise, highlight the many pressures and forces at work that adamantly remain at odds with a more enlightened stance of sensitivity and understanding. The intent of the "Don't Ask, Don't Tell" policy is to allow quiet and celibate homosexuals to serve in the military. Under Clinton's policy, there is no "duty to report" for psychologists or other health care providers. All members of DoD have a "duty to report" in cases of violation of law or military regulations. Quiet and celibate active-duty homosexuals violate neither the law nor military regulations. For example, as the authors of this chapter understand the facts of the Blaesing case, the psychologist had no "duty to report" the corporal's concerns that he might be homosexual.

Conclusions and Recommendations

Our country was founded on the need to preserve individual rights within the context of a free society. Throughout our history, legal decisions and statutes have sought to maintain individual freedoms such as autonomy, self-determination, and privacy, as have the standards set for appropriate clinical care and the ethical principles that guide the decision-making and judgment of health care providers.

Doctor-patient, therapist-patient, and clergy-parishioner privilege have been granted legal status as acknowledgment of the need for a confidential relationship so that the trust necessary for effective treatment and counseling may be established. As Weiner and Wettstein (1993) stated, "confidentiality is considered the *sine qua non* of successful psychiatric treatment. Without the promise of confidentiality, provided primarily through the clinician's professional ethics but also the law, many individuals in need of treatment would be afraid to seek it. It is even clearer that once in treatment, clients would be affected by the absence of confidentiality" (p. 202).

The absence of confidentiality is deleterious to the therapeutic process. Without it, individuals are unlikely to enter treatment and the value of such services may be severely limited. Moreover, as highlighted in the examples presented, breaches of confidentiality can lead to grave consequences.

Military regulations specifically allow for a variety of privacy invasions and breaches of confidentiality under the guise of need to know. Individual rights are greatly limited and the powers entrusted to DoD officials are extensive. All consumers of counseling services within DoD should understand this.

As noted in Jeffrey, Rankin, and Jeffrey (1992), the psychology consultant to the surgeon general of the Army Medical Department stated that confidentiality did not exist for consumers treated by military psychologists (Fishburne, 1987). Moreover, military spokespersons consistently have affirmed that the DoD does not recognize patient confidentiality (Sarhaddi, 1989; Raddatz, 1994). Although such pronouncements may sound extreme, they reflect constraints placed on military health care providers.

The Code of Ethics of the American Psychological Association (APA, 1992) requires psychologists not to disclose patient information without the patient's consent. DoD regulations require the release of such information to the chain of command without requiring the patient's consent or knowledge when a need to know exists. The need to know is not a clearly defined concept, and this criterion may be used arbitrarily. Thus, no real confidentiality exists for military service members. Military psychologists who fail to understand this risk reprimand by the American Psychological Association (Jeffrey, 1991).

Homosexual military personnel should have the same right to high-quality medical, mental health, and spiritual care as their heterosexual counterparts. Although all service members must be aware of the restrictions placed on confidentiality in the military setting, homosexual members must be especially aware of the potential limits to confidentiality and the possible misuses of information that might be obtained in a therapeutic relationship.

All homosexual service members should be educated about these issues. Health care providers should advise each consumer of the limits to confidentiality that may exist. This should be done in writing and be a part of each consumer's consent to treatment or evaluation. It should include an explanation of the types of information that may be released and the possible uses of such information (for an example, see the appendix). It is imperative that this disclosure be made at the outset of any treatment or evaluation relationship so that service members' rights may be preserved. It is crucial that service members be given adequate information so as to make informed decisions both about entering health care relationships and about how much information it would be prudent to share in these relationships.

In summary, despite the long precedents of doctor-patient, therapist-patient, and clergy-parishioner confidentiality, and the fact that all states have statutes guaranteeing this much-needed principle, military regulations usurp these statutes as well as professional ethics codes in the name of national security. Confidentiality does not exist within the military. Contrary to expectations, the military and its representatives are the holders of the privilege, not the service members themselves. So that each individual's rights, dignity, and worth may be preserved, and so that their trust will not be exploited, it is crucial that confidentiality not be promised. Limits of confidentiality should be disclosed in advance. Expectations for all therapy and evaluation relationships must be delineated and understood in advance, and consent should be in writing so that adverse consequences will not befall any service member. In the absence of the more complete confidentiality provided by health care providers and clergy in the private sector, all service members and their families should be provided the above.

References

American Psychiatric Association. (1952). *Diagnostic and statistical manual: Mental disorders (DSM-I)*. Washington, DC: Author.

———. (1987). Guidelines on confidentiality. *American Journal of Psychiatry, 144*, 1522–1526.

———. (1994). *Diagnostic and statistical manual of mental disorders* (4th ed.). Washington, DC: Author.

American Psychological Association. (1987). *General guidelines for providers of psychological services*. Washington, DC: Author.

———. (1992). Ethical principles of psychologists and code of conduct. *American Psychologist, 47* (12), 1597–1611.

Barefoot v. Estelle, 463 U.S. 880 (1983).

Cohen, R. J., & Mariano, W. E. (1982). *Legal guidebook in mental health.* New York: Free Press.

Confidentiality of Medical Records, Md. Code Ann., Health–General, subtit. 3, §§4-301 through 4-403 (1991).

Creswell, C. E. (1981). Privileged communication in the military chaplain. *Military Chaplain's Review,* DA Pamphlet 165-128, pp. 7–16.

Denham, T. E., & Denham, M. L. (1986). Avoiding malpractice suits in pastoral counseling. *Pastoral Psychology, 35,* 83–93.

Department of Defense (1975). *Personal privacy and the rights of individuals regarding their personal rights.* (DoD Directive 5400.11). Washington, DC: U.S. Government Printing Office.

———. (1980). *DoD Freedom of Information Act Program.* (DoD Directive 5400.7). Washington, DC: U.S. Government Printing Office.

———. (1985). *Licensure of DoD health care providers.* (DoD Directive 6025.6). Washington, DC: U.S. Government Printing Office.

———. (1994). *Manual for courts-martial, United States.* Washington, DC: U.S. Government Printing Office.

Department of the Army (1981). *Alcohol and drug abuse prevention and control program.* (Army Regulation 600-85). Washington, DC: U.S. Government Printing Office.

———. (1982a). *Personnel separations: Officer personnel.* (Army Regulation 635-100). Washington, DC: U.S. Government Printing Office.

———. (1982b). *Release of information and records from Army files.* (Army Regulation 340-17). Washington, DC: U.S. Government Printing Office.

———. (1985). *The Army privacy program.* (Army Regulation 340-21). Washington, DC: U.S. Government Printing Office.

———. (1987). *Medical record and quality assurance administration.* (Army Regulation 40-66). Washington, DC: U.S. Government Printing Office.

Doe v. Roe, 93 Misc. 2d 201, 400 N.Y.S. 2d 688 (Sup. Ct. 1977).

Dyer, A. R. (1988). *Ethics and psychiatry.* Washington, DC: American Psychiatric Press.

Fishburne, F. J., Jr. (1987, August). *Army medical department psychology business meeting.* Discussion at the annual meeting of the American Psychological Association, New York.

Griswold v. Connecticut, 381 U.S. 479 (1965).

Gutheil, T. G., & Appelbaum, P. S. (1982). *Clinical handbook of psychiatry and law.* New York: McGraw-Hill.

Hastings Center. (1978). *In the service of the state: The psychiatrist as double agent* (special supplement). Hastings-On-Hudson, NY: Author.

Howe, E. G. (1989). Confidentiality in the military. *Behavioral Sciences and the Law,* 31–33.

Jacob, M. R. (1982). Confidentiality: Re-examined and reapplied. *Military Chaplains' Review,* DA Pamphlet 165-134, pp. 31–38.

Jeffrey, T. B. (1989). Issues regarding confidentiality for military psychologists. *Military Psychology, 1,* 49–56.

———. (1991). The Army vs. APA: Conflicting patient confidentiality requirements. *The Military Psychologist, 8,* 11–12.

Jeffrey, T. B., Rankin, R. J., & Jeffrey, L. K. (1992). In service of two masters: The ethical-legal dilemma faced by military psychologists. *Professional Psychology: Research and Practice, 23,* 91–95.

Lipari v. Sears, Roebuck & Co., 497 F. Supp. 185 (D. Neb. 1980).

McIntosh v. Milano, 168 N.J. Super. 466, 403 A.2d 500 (1979).

Meyer, R. G., Landis, E. R., & Hays, J. R. (1988). *Law for the psychotherapist.* New York: Norton.

Miller, D. J., & Thelen, M. H. (1986). Knowledge and beliefs about confidentiality in psychotherapy. *Professional Psychology: Research and Practice, 17,* 15–19.

Morra v. State Board of Examiners of Psychologists, 212 Kan. 103, 510 P.2d 614 (1973).

New York Evidence Law §4504(a) (McKinney 1962).

Pope, K. S. (1990). A practitioner's guide to confidentiality and privilege: 20 legal, ethical, and clinical pitfalls. *The Independent Practitioner, 10,* 40–44.

Pope, K. S., & Vasquez, M. J. T. (1991). *Ethics in psychotherapy and counseling: A practical guide for psychologists.* San Francisco: Jossey-Bass.

Privacy Act, 5 U.S.C. §552a (1974).

Raddatz, M. (1994, June 8). *Morning Edition.* National Public Radio. Transcript #1363-11.

Reisner, R. (1985). *Law and the mental health system: Civil and criminal aspects.* St. Paul, MN: West.

Roe v. Wade, 410 U.S. 113 (1973).

Sarhaddi, S. (1989, August 26). Therapist ordered to testify. *The (Middletown, N.Y.) Times Herald Record,* pp. 3, 12.

Shah, S. A. (1969). Privileged communications, confidentiality, and privacy: Privileged communications. *Professional Psychology, 1,* 56–59.

Sharkin, B. S. (1995). Strains on confidentiality in college-student psychotherapy: Entangled therapeutic relationships, incidental encounters, and third-party inquiries. *Professional Psychology: Research and Practice, 26,* 184–189.

Simon, R. I. (1992). *Clinical psychiatry and the law.* Washington, DC: American Psychiatric Press.

Stromberg, C. D., Haggarty, D. J., Leibenluft, R. F., McMillian, M. H., Mishkin, B., & Rubin, B. L. (1988). *The psychologist's legal handbook.* Washington, DC: Council for the National Register of Health Service Providers in Psychology.

Tarasoff v. Regents of University of California, 17 Cal. 3d 425, 551 P.2d 334, 131 Cal. Rptr. 14 (1976).

Thompson v. County of Alameda, 614 P.2d 728 (Cal. 1980).

Trower, R. H. (1992). The chaplain as counselor. In R. J. Wicks, R. D. Parsons, and D. Capps (Eds.). *Clinical handbook of pastoral counseling* (Vol. 1). New York: Paulist Press.

Vassiliades v. Garfinckels, 492 A.2d 580 (D.C. Ct. App. 1985).

Warren, S., & Brandeis, L. D. (1890). The right to privacy. *Harvard Law Review*, 4, 193.

Weiner, B. A., & Wettstein, R. M. (1993). *Legal issues in mental health care.* New York: Plenum Press.

Wicks, R. J., Parsons, R. D., & Capps, D. (1993). *Clinical handbook of pastoral counseling.* (Vol. 1). New York: Paulist Press.

Appendix

Limits of Confidentiality

It is important for you to know the limits of confidentiality of psychological information. Department of Defense (DoD) medical records are the property of the Government, thus the same controls that apply to other Government documents apply to them. Information disclosed by patients to DoD Medical Department health personnel is not privileged communication.

Access to information in your medical record is allowed when required by law, regulation, or judicial proceedings; when needed for hospital accreditation; or when authorized by you.

Examples of the limits of confidentiality follow:

1. If a provider of health care services believes you intend to harm yourself or someone else, it may be the duty of the provider to disclose that information.
2. In situations of suspected child abuse, it is the duty of the provider to notify medical, legal, or other authorities.
3. If you are involved in any legal action/proceedings your records may be subject to subpoena.
4. Other members of the professional staff associated with your health care may have access to information on record without your written consent.
5. DoD officials could have access to information authorized by regulation, e.g., a command directed referral, a line of duty investigation, or participation in the nuclear surety program.
6. Qualified persons may have access to your record for clinical investigation purposes.

If you have questions about the limits of confidentiality you may ask us or inquire at the Patient Administration Division of the hospital.

Statement of Understanding

I have read the above and understand that information about me will be safeguarded within the limitations of confidentiality mentioned above and in the Privacy Act Statement (DD Form 2005).

_____   _____

Patient Signature   Date

Implementing Policy Changes in Large
Organizations: The Case of Gays
and Lesbians in the Military

Gail L. Zellman

Whatever its form or content, any policy that allows acknowledged homosexuals to serve in the military has to be implemented in an organization that, like most organizations, resists changes in those structures, policies, and practices that have contributed to its past success. Even though military organizations are accustomed to rapid changes in technology and battle threats, they are usually highly averse to *social* changes—that is, changes in their traditions, customs, and culture (Builder, 1989).

In the case of allowing acknowledged homosexuals to serve in the military, the resistance to change touches not only on deeply held attitudes, but, for a large portion of the military, on moral beliefs as well. For many, it makes no difference if a service member ever comes in contact with an acknowledged homosexual: The change in policy itself alters their perception of *their* organization in a fundamental way.

This chapter considers how such a policy might be effectively implemented, in light of institutional culture, the recent history of efforts to change policies toward gays, the current policy context, and what the literature tells us about implementing policy change in large organizations. The chapter first discusses salient features of military culture and the current policy context. Then, it reviews factors that constrain and support policy implementation, including policy design, features of the

Much of the material in this chapter was drawn from an earlier chapter prepared by Gail L. Zellman, Joanna Zorn Heilbrunn, Conrad Schmidt, and Carl Builder (1993).

implementation process, and the local context for change. It then describes efforts to change military policy toward gays and lesbians that were initiated in the early days of the Clinton administration. Current policies concerning the inclusion of gays and lesbians in the military are next presented. The chapter ends with a discussion of the importance of implementation in bringing about organizational change, and the perils of ignoring its importance.

Implementation Context

Implementation as an area of study resulted from a need to understand why policy changes imposed from the top of large organizations often failed to appear at the bottom, or, if they did, why they looked so different from what their framers had intended. These findings challenged widespread beliefs that organizational change is a relatively straightforward process with predictable outcomes.

The literature on the implementation of innovations in large organizations focuses heavily on the introduction of technological or organizational change (e.g., O'Toole, 1989; Langbein and Kerwin, 1985; Prottas, 1984; Wilms, 1982; Zetka, 1991; Walsh, 1991). Social changes, which inherently involve much more deeply held attitudes about race, religion, sexuality, or values, bring added complexity to the change process. Externally imposed social change challenges an organization and its leadership to create a blueprint for change that considers the institutional culture and incorporates useful implementation theory principles, a large measure of leadership, an understanding of the extent to which previous experience applies, and a keen awareness of the fears and limits of those at the bottom, on whom the success of policy implementation ultimately depends.

Military Culture
The military can be described as an organization that is based on a formal, hierarchical, and rule-driven structure, which values efficiency, predictability, and stability in operations. This structure is supported and reinforced by conservative organizational and participant cultures. Many of the values of military families still reflect those of small towns and of several decades past, which may reflect the selective enlistment inherent in the all-volunteer force. For many of the more senior military people now in leadership positions, there remains a legacy of the hostility be-

tween the military and the rest of society that reached a peak during the war in Vietnam. For these people, the imposing of unwelcomed aspects of society on the military—often referred to as "social experimentation"—evokes familiar and hostile feelings.

The Policy Context

The military has seemed particularly averse to removing the restriction on homosexuals because that policy would threaten its cultural values and would be externally imposed. Five factors make the integration of homosexuals particularly problematic. These factors clearly emerged in focus groups conducted by study staff at military bases in the United States and Germany (Berry, Hawes-Dawson, and Kahan, 1993).

First, a majority of military personnel, and a sizable portion of the general public, believe that homosexuality is immoral. For many, allowing homosexuals to serve would put the military in the position of appearing to condone a homosexual lifestyle.

Second, the debate on whether to allow acknowledged homosexuals to serve in the military began and has continued in a context characterized by drawdowns (reductions in the size of the military) and uncertainty. In response to the end of the Cold War, the military's role and mission are being widely questioned. Reduced military budgets have created considerable anxiety among military personnel. Many believe that with base closings, drawdowns, and reductions in benefits, the military has violated the psychological contract between the organization and its members (Rousseau, 1989). The resulting anger and resentment have made members disinclined to tolerate additional threats to military culture.

Third, the policy debate is occurring in a context in which norms of deference are significantly eroded. Military members and leaders have outspokenly and harshly criticized the commander in chief and his policies. Outspoken opposition to permitting homosexuals to serve is a cause for concern because it sends the message that a new policy would have no support among top military leaders. Nevertheless, norms of obedience remain and some observers argue that these norms will carry the day. Indeed, on June 10, 1993, in a speech at Harvard University, the chairman of the Joint Chiefs of Staff, General Colin Powell, said of a new policy toward homosexuals' military service, "The President has given us clear direction. . . . Whatever is decided, I can assure you that the decision will be faithfully executed to the very best of our ability."

Fourth, the current budgetary context may restrain change if implementation planning fails to take it into account. Because budgets are not growing, all new programs come at the expense of old and sometimes cherished ones. We can expect that the more that any integration process costs, the more resentment it will elicit.

Fifth, there is no sense that the change would serve any legitimate need of the military. Objections that the policy is not based on need are reinforced by the sense among many military members that even the president is not committed to the change. The president's retreat from his initial policy reinforced this belief.

Military structure and culture and key features of the policy context are unique to the problems of implementing a policy to allow homosexuals to serve. Yet empirical findings and general principles derived from studies of policy implementation and organizational change offer lessons for implementing such a policy. These literatures and the lessons that they offer are described below.

Factors That Constrain and Support Policy Implementation

Implementation itself is best defined as "the carrying out of a basic policy decision, usually incorporated in a statute but which can also take the form of important executive orders or court decisions. Ideally, that decision identifies the problem(s) to be addressed, stipulates the objective(s) to be pursued, and in a variety of ways, structures the implementation process" (Mazmanian and Sabatier, 1983, p. 20). Policy analysts often divide the change process into two phases: adoption and implementation. The adoption phase begins with the formulation of a new policy proposal and ends when that proposal is formally encoded in a law, regulation, or directive. The implementation phase begins with the formal adoption of the policy and continues at some level as long as the policy remains in effect (e.g., Weimer and Vining, 1992).

Those who study implementation generally agree that three categories of variables contribute most significantly to policy change: (1) policy design; (2) the nature of the implementation process; and (3) the local organizational context in which the policy is implemented (e.g., Mazmanian and Sabatier, 1983; Goggin, 1987). Each of these components is discussed in turn.

Policy Design
The design of a new policy and its expression in a policy instrument can substantially affect both the implementation process and the extent to

which the policy's original objectives are met in practice. Those policy design components that most affect outcomes include characteristics of the change required and the nature of the policy instrument.

Characteristics of the required change. Some changes are inherently more complex than others. For example, a law designed to reduce highway fatalities by lowering the speed limit contains within itself all the information necessary to enable individuals to comply (McDonnell and Elmore, 1987). In contrast, a court order to create equal educational opportunity is less clear. Individuals must not only read and understand the equality standard, but also create a plan that translates the goal into required behaviors, a much more complex task (McDonnell and Elmore, 1987).

A policy's successful implementation also derives from the validity of the causal theory that underlies it. Every major reform contains, at least implicitly, a causal theory linking prescribed actions or interventions to policy objectives. To the degree that there is consensus about the validity of the theory (i.e., that most agree that by carrying out the intervention, attainment of policy objectives is likely), policy implementation is facilitated (Mazmanian and Sabatier, 1983).

Another key characteristic of the required change is the scope of change required. Scope can be measured in terms of the size of the target group, the percentage of the population affected, or the extent of behavior change required. In general, policies that require less change, in terms of numbers and extent, are easier to implement (Mazmanian and Sabatier, 1983).

Nature of the policy instrument. McDonnell and Elmore (1987) describe four generic classes of policy instruments: (1) mandates, which are rules governing the actions of individuals and agencies, intended to produce compliance; (2) inducements, the transfer of funds to individuals or agencies in return for certain agreed-upon actions; (3) capacity-building, the transfer of funds for investment in material, intellectual, or human resources; and (4) system-changing, the transfer of official authority among individuals and agencies to change the system through which public goals and services are delivered.

The choice of instrument structures affects the implementation process to a significant degree. Expected outcomes, costs, and the extent of oversight all vary by type of policy instrument. For example, whereas mandates seek uniform but minimal compliance, inducements are designed to produce substantial variability in outcomes because there is often a variety of ways to achieve high performance. Mandates require a strong focus on coercion and compliance monitoring, whereas the implementation of inducements requires oversight but no coercion (McDonnell and Elmore, 1987).

Implementation guidance. Implementation guidance is built into some policies, such as a reduced speed limit, as noted above. In other cases, guidance is less inherent, but may be included in several forms. Among the most important ways to include guidance are clearly ranking policy objectives and stipulating decision rules for implementing agencies.

A clear ranking of policy objectives is indispensable for program evaluation and for directing the actions of implementing officials. Statements about objectives may also be used as a resource for groups that support the policy objectives. Formal decision rules of implementing agencies, such as the stipulation in a statute of the level of support required for a specific action (e.g., two-thirds majority of a specified commission required for a license to be issued), reduce ambiguity and increase the likelihood that a mandate will be carried out as intended (Mazmanian and Sabatier, 1983).

Implementation Process
Implementation researchers (e.g., Elmore, 1978, 1980; Goggin, 1987; McLaughlin, 1987, 1990; Mazmanian and Sabatier, 1983) view the process through which a new policy is implemented as a key contributor to understanding organizational change. From the implementation perspective, any analysis of policy choices or the effects of policy on organizations matters little if implementation is poorly understood.

Several key notions emerged from the early implementation studies (McLaughlin, 1990). First, changing practice through policy is a difficult proposition. Second, policymakers cannot mandate what matters—capacity and motivation at the lower levels of the organization, where the policy must find a home. Third, by focusing on policy implementation, certain processes and rules could be utilized to increase the likelihood that policy would be implemented, relatively unchanged, into practice (Mazmanian and Sabatier, 1981).

These notions suggest an implementation process structured around pressure and support. Pressure, McLaughlin (1987) argues, focuses attention on the new policy and increases the likelihood of compliance; support is necessary to enable implementation. Such support may include adequate financial resources, a system of rewards that recognize compliance efforts, and room for bottom-level input into the process.

Any policy change with regard to homosexuals serving in the military will be mandated. The implementation of a mandate necessarily involves top-down processes, although the considerable discretion accorded lower-level military leaders means that their support of, or opposition to, a new policy is key to its success.

Research on regulatory policy has demonstrated that targets of mandates incur costs from complying or from avoiding compliance. The choice they make to comply with the mandate or attempt to avoid doing so is based on the perceived costs of each alternative. Targets decide whether to comply by calculating two kinds of costs: (1) the likelihood that the policy will be strictly enforced and compliance failures will be detected; and (2) the severity of sanctions for noncompliance. If enforcement is strict and sanction costs are high, compliance is more likely (McDonnell and Elmore, 1987).

Pressures. To increase the likelihood of compliance with a mandate, the implementation plan must include enforcement mechanisms and sanctions that lead targets to assess the costs of noncompliance as high, and thus increase the likelihood that they will choose to comply. But such a plan is likely to create an adversarial relationship between initiators and targets, particularly when targets do not support policy goals (McDonnell and Elmore, 1987).

Support. Along with pressure to comply, policy mandates should provide support for implementation. A set of rewards for any movement that supports implementation of the policy is key. The goal of these rewards is for individuals to perceive that their self-interest lies in supporting the change. Such beliefs represent the energizing force for successful implementation of change (Mazmanian and Sabatier, 1983; Levin and Ferman, 1986).

Mazmanian and Sabatier (1983) note the importance of committed implementors as driving forces for policy change. Conversely, leaders uncommitted to a new policy may restrain change efforts. Indeed, the authors suggest that the inability of policymakers or organizational leaders to choose implementors is a major factor in implementation failures. If implementors cannot be replaced—and often they cannot—the leader's job is to change the perceptions of the implementors concerning the likely personal outcomes of the new policy. If implementors come to view the new policy as consistent with their self-interest (Mazmanian and Sabatier, 1983) and with organizational culture (Schein, 1987), they will be far more likely to support the new policy and act in ways that enhance its implementation.

Local Context for Change

When an organization's culture appears inconsistent with a new policy, as would be the case with any policy that allows homosexuals to serve in the military, leaders must attempt to create driving forces for change by drawing on those aspects of the existing culture that *are* compatible (Allaire and Firsirotu, 1985; Schein, 1987).

A new policy is most likely to *clash* with organizational or participant culture when it is imposed from the outside, a common occurrence in government agencies. In such cases, the new policy may reflect the demands of outside constituencies, the results of research, or the opinions of policymakers.

Often, an externally imposed policy threatens the premium put on history and learning from experience in the organization (Schein, 1987; Levitt and March, 1988). In some cases, such policy changes are perceived to threaten the organization's very survival. The policy can also threaten deeply held beliefs concerning organizational autonomy, a key feature in the widespread resistance of school districts to desegregation orders.

Change may be inconsistent with organizational *structure* as well as culture. Allaire and Firsirotu (1985) note that innovations that depend on a particular organizational structure are likely to fail if those structures do not exist in the organization. For example, it would be futile, they argue, to exhort the employees of a regulated monopoly offering a public service and requiring large capital investments to manage with a lean staff and simple form.

A key finding of implementation studies is that change is best accepted and institutionalized when at least some people within the organization perceive the need for the change and are persuaded that it is good for the organization and for themselves. Change imposed from without lacks these built-in advantages. The process of change must be much more carefully planned and managed if widespread implementation, consistent with policy goals and processes, is to occur. Even when policy, culture, and structure are consistent, implementation is far from assured. The natural conservatizing forces at work in most organizations tend to resist change. People often have to be persuaded that the new policy will not be harmful to the organization or to themselves, and that it may even result in gains.

Implementing a Policy to End Discrimination

How might the military have implemented a policy similar to the one that RAND (Zellman et al., 1993) identified as potentially workable— one that sought to end discrimination against homosexuals by considering sexual orientation, by itself, as "not germane" to determining who may serve in the military, and that was based on clear standards of conduct, strictly enforced? The nature of military organizations and our knowledge about the implementation process suggest a number of actions that would facilitate the implementation of such a policy. These actions, described below, are discussed in more detail in the RAND report (Zellman et al., 1993).

Design a Policy That Facilitates Implementation

It is very important to convey a new policy that ends discrimination as simply as possible and to impose the minimum of changes on personnel (Levin and Ferman, 1986). Moreover, the policy should be decided upon and implemented as quickly as possible and should include both pressure and support for change.

Make the policy simple. Military experience with African Americans and women (see chapters 4 and 5) argues for a simple policy under which homosexuals are treated no differently in terms of work assignments, living situations, or promotability.

Much can be learned from the military's experiences in integrating women. The policy message about military women has been complex. This complexity has resulted in continuing strong doubts about the capability and appropriateness of women to perform certain tasks. Lower training standards, better assignments (to safer, noncombat jobs), and better accommodations (largely because the small numbers of women lower ratios for toilets, etc.) have continued to cause resentments among men. These problems in integrating women argue for equal treatment of homosexuals. They should be assigned to serve in all positions and at all levels, according to their skills; those who serve with them will be expected to treat them equally as well. Indeed, the documented capabilities of homosexuals to perform all military tasks enable the policy to be simple.

Act quickly. Lessons from the implementation literature suggest that a new policy regarding homosexuals in the military should be decided upon and implemented as quickly as possible, for three reasons. First, the waiting period is one in which personnel are unsure, and therefore anxious about, what the final outcome will be and how it will affect them. In fact, the change in policy will not appreciably affect the vast majority of heterosexuals, who will not be working or living with an open homosexual. (See chapter 8 for a discussion of the probabilities of acknowledged homosexuals in groups of varying sizes.) Once they discover that nothing has changed for them, they will feel more comfortable and the issue will be less disruptive generally. That outcome, however, will require that instances of open homosexuality not be allowed to result in serious, rumor-inspiring conflicts.

Second, any waiting period permits restraining forces to consolidate. Until the policy is decided upon and implementation has begun, those opposed will feel free to speak out against it, increasing the perceived strength of military objections.

Third, fast and pervasive change will signal commitment to the policy. Phased-in implementation might allow enemies of the new policy intentionally to create problems to prove the policy unworkable.

Convey the change in terms compatible with military culture. To the extent possible, the policy should be conveyed in terms compatible with military culture. For example, leadership should focus on the organizational culture of hierarchy and obedience and minimize discussion of the inconsistency between the new policy and a very conservative participant culture. Leaders can become role models by conforming behaviorally to the new policy because the president is the commander in chief, who must be obeyed. Other consistencies between successful implementation of the policy and organizational culture can be stressed. For example, the military sees itself as a strong organization with a "can-do" attitude. Military culture stresses the dominance of mission over individual preferences and characteristics. Such successful submersion of more visible differences such as race can be pointed to as an example of the military's ability to keep its eye on the prize. The military's norms of inclusion and equality can be effectively utilized as well.

Build in sanctions and enforcement mechanisms. Any new policy about homosexuals will be mandated. Consequently, compliance is the goal. To increase the likelihood of compliance, sanctions and enforcement mechanisms must be established.

Crucial to promoting compliance is the adoption or revision of a code of professional behavior that clarifies the criteria for behavioral compliance on the part of both homosexuals and heterosexuals. The code must include some general principles, general behavioral criteria, and some language that explicitly makes people responsible for exercising discretion in determining whether behaviors not explicitly included in the code are acceptable (Burke, 1990). The code should explicitly recognize the need to respect the feelings and concerns of others in defining acceptable and unacceptable behaviors.

Although the military's strong hierarchical control might suggest to some that policy can be successfully implemented with only limited discretion (Burke, 1990), providing some degree of discretion to the smallest unit in terms of how to bring about behavior change captures an important tenet of the implementation perspective. Indeed, the military mission order, a widely used way of directing subordinates, allows considerable lower-level discretion. Such discretion increases individual and unit commitment to the change.

The code of professional conduct must also describe the sanctions for noncompliance. These sanctions essentially define accountability and thus set parameters around leader discretion. Too much discretion concerning sanctions risks the possibility that uncommitted leaders will send a signal that inappropriate behavior will be tolerated. This happened in the Reagan administration, when political appointees who opposed environmental laws headed the Environmental Protection Agency (EPA). "The atmosphere created by Reagan appointees who headed EPA discouraged civil servants from serious enforcement of social environmental laws. They were encouraged to use their discretion to reduce the scope of effective enforcement" (Palumbo and Calista, 1990, p. 8).

The enforcement system must be made explicit (Elmore, 1987). Organization members must understand that their behavior will be observed and noted and that actions inconsistent with the code of behavior will be reported to their superiors and penalized according to the explicit sanction policy. But military experience in the area of sexual harassment demonstrates that a code of professional conduct *by itself* is not enough to ensure change when the change is inconsistent with organizational culture. Rather, it must be just one part of an intensive implementation effort if change is to occur.

Ensure Leadership Support at All Levels

To effect change, military leaders must be dynamic and forceful in promoting it. They assume this role when they are perceived to be supportive of the change and to be concerned that it will be successfully implemented. Such a position is sometimes difficult to achieve, especially when the new policy has been criticized by these same leaders early in the process, when debate was occurring about the policy's value and form. Ideally, the early criticisms of leaders are acknowledged and responded to during the policy formulation process in a way that enables them to emerge from the debate appearing convinced of the value and importance of the new policy (Allaire and Firsirotu, 1985).

If lower-level commanders and troops do not believe that their superiors support the policy, they will have little motivation to abide by it. The commander in chief must reaffirm his commitment to a new policy in language consistent with cultural norms of inclusion and equality for all. If senior military leaders do not believe in the change, efforts must be made to present leaders as *behaviorally* committed to the policy (even if they remain attitudinally opposed).

Such behavioral commitment requires that leaders send a strong, consistent signal of support for the new policy. Lack of attitudinal support makes behavioral signaling all the more important. Such signaling must include strict adherence to an existing or new code of professional conduct, with public sanctioning of personnel at all levels who fail to comply with it. It must also include smaller actions, such as continuing repetition of support (Peters, 1978). The assignment of a high-ranking individual with direct access to top management to oversee the implementation process conveys the message that this policy is to be enforced at all levels.

Although top-down change is the norm in military organizations, it is important to convey an understanding of what matters at the bottom of the organization so that members believe that their concerns are heard. It is important, as well, to convince leaders at all levels, including the bottom, that it is in their and the organization's interest to work to support the new policy. Their effective involvement depends on six key efforts: (1) signaling the military's commitment to the new policy; (2) convincing them that active monitoring and support for the new policy will be noticed and rewarded, and that breaches of policy by subordinates will be considered instances of leadership failure; (3) stressing the importance of reducing anxieties and creating a sense of perceived fairness for members; (4) training them to be good implementors; (5) empowering them to use their discretion within clear constraints; and (6) providing guidance.

Signal commitment. Lower-level leaders are the key to enforcement efforts at the bottom of the military hierarchy. Unless the seriousness of the military's commitment to the policy is effectively conveyed to them, they will exhibit great variability in their enforcement efforts. In the absence of a strong message of conformity from superior officers, treatment of the same issue can be expected to differ considerably from base to base, and unit to unit.

Identify rewards. The enforcement system must be made explicit (Elmore, 1978). Leaders must be persuaded that their enforcement of the new policy will be monitored by those above them and that their behavioral support of the new policy will be rewarded. This will encourage leaders to believe that successful implementation of the new policy accords with their own self-interest, a key aspect of leadership (Levin and Ferman, 1986).

Communication upward about compliance failures should be actively encouraged (Dalziel and Schoonover, 1988). Because "snitching" violates a tenet of military culture that only good news should be communicated, it is important to redefine "snitching" as important, valued professional behavior and to set up monitoring procedures so that people are *asked* about problems, for example, through regular implementation surveys (e.g., Gottlieb et al., 1992).

Leaders must also understand that failure to actively support the new policy will be noticed and sanctioned. Both heterosexual and homosexual military members must be held to high standards of conduct with regard to abiding by and enforcing the new policy. Any officer who violates the behavioral guidelines associated with the new policy should be disciplined severely. This message—that the military takes the new policy seriously—will quickly be conveyed to subordinates and contribute to behavioral compliance.

Moreover, breaches of policy by subordinates must be viewed and treated as leadership failures. This dual approach makes every leader responsible for the behavior of subordinates. More generally, commanders must be responsible for morale and behavior within their units, including all incidents of discrimination. It must be made clear to them that if they permit an environment in which homosexuals can be discriminated against or harassed, it will have an effect on their likelihood of promotion. Failure to pursue instances of unacceptable behavior should, in itself, be considered a leadership failure. This latter point is key: Perceptions about what happens when these responsibilities are ignored can enhance or inhibit implementation (M. Davidson, American Civil Liberties Union attorney, personal communication, May 18, 1993).

The implementation leader must clarify the complaint process and ensure that complaints are actively addressed. Efforts should also be made to simplify the complaint process as much as possible. The Army Equal Opportunity Office is implementing two promising approaches: (1) a hot line that provides procedural information on filing equal opportunity complaints; and (2) a complaint form that can be reproduced easily on a photocopier (K. Clement, U.S. Army Equal Opportunity officer, personal communication, April 29, 1993).

Strengthen the Local Context for Change

Change will be facilitated by leaders who are trained and motivated to address and solve implementation problems. A new organizational structure should be helpful as well in enabling implementation and change. Finally, monitoring criteria should be developed and widely communicated.

Increase leadership capacity. A key task of leaders at all levels is to minimize anxieties and create a sense of procedural justice for subordinates. Reduced worry and feelings of justice are enhanced when leaders are prepared to absorb the anxiety of change, including challenges and anger, when leaders demonstrate dedication and commitment to the organization as a whole, when leaders encourage members to express their anxieties and concerns, and when leaders acknowledge members' concerns (Schein, 1987; Tyler and Lind, 1992).

Leaders should also act to enhance feelings of efficacy by conveying their beliefs that personnel are capable of implementing the change and conforming to behavioral expectations. The critical distinction between behavior change and attitude change should be emphasized, with a clear message that the organization will limit its concern exclusively to behavior.

Leadership capacity must be enhanced by several means, including training, support for the use of discretion, and guidance.

Conduct training. Training of leaders should be designed to create "fixers"—people who both care about successful implementation and have the skills necessary to anticipate and identify implementation problems and to make adjustments to improve the implementation process (Bardach, 1980; Levin and Ferman, 1986).

It should be noted that "fixer training" is distinctly different from sensitivity training. Fixer training is practical and complements the strictly behavioral approach to implementation most likely to yield success. In contrast, sensitivity training attempts attitude change and is widely scorned by military personnel (similar to attitudes of police and firefighters described in chapter 7). Bringing in sensitivity trainers who are perceived to be very costly in a context of drawdown is as likely to increase resistance and anger as it is to reduce it.

Encourage use of discretion. Becoming a good "fixer" implies the possibility of action. Leaders at all levels must be accorded sufficient discretion so that they can act to correct implementation problems. But as noted above, this discretion must be bounded by behavioral monitoring and strict enforcement of a code of professional conduct.

Provide guidance. Any code of professional conduct, no matter how prescriptive, cannot hope to identify all potential problem areas. A new code of professional conduct that describes behavioral principles and goals will identify few. Yet lower-level leaders need guidance. Therefore, codes should be supplemented with active guidance in the form of "question and answer" documents, which should be widely disseminated. These questions and answers could also educate by including information about sexual behavior and health issues.

Monitor compliance. Monitoring should be carried out by using the chain of command. Monitoring should begin among low-level leaders. They should report on a periodic basis to their superiors and should be provided incentives, as described above, to report in a timely manner about incipient problems so that they can be remedied before they become serious. Such reporting up the chain of command will depend upon the development of clear reporting instruments, and on creating among leaders at all levels a sense that accurate information about implementation problems is valued and that failures of leadership reside in refusals to comply, not in compliance difficulties.

This process should be supported by a small group in each service charged with overseeing implementation of the new policy. The group may comprise people already responsible for other similar policies, such as sexual harassment and racial integration.

Develop monitoring criteria. Few homosexuals are likely to reveal their sexual orientation even if a policy that allows them to do so openly is mandated. Consequently, monitoring criteria such as numbers of promotions, distribution across pay grades, and other measures of a group's progress that depend on the ability to detect group numbers are not feasible.

But it is possible and important to monitor other outcomes of the implementation process. These outcomes should include key areas of concern, including incidents of violence, numbers of open homosexuals who serve, and measures of unit performance.

Surveys of member attitudes toward the new policy and experiences with it could be a valuable monitoring device, particularly if the surveys occurred frequently, perhaps at six-month intervals. A set of questions focused on the implementation of the new policy toward homosexuals would allow the monitoring group to examine key issues, such as behavioral compliance, reporting behaviors, and (for commanders), the extent to which implementation of the policy coincided with other duties (Gottlieb et al., 1992). The opportunity to track implementation over time through a mix of unchanging attitudinal and changing implementation questions would be invaluable.

Nonimplementation and a Policy Shift

The "not germane" policy proposed as potentially workable (National Defense Research Institute, 1993) was never implemented. Indeed, President Clinton backed down in the face of considerable opposition to lifting the ban on homosexuals by the uniformed military and some members of Congress.

Clinton's failure to stand by his pledge to lift the ban, regardless of one's views concerning its political wisdom, complicated implementation of any policy whose goal is to encourage or permit homosexuals to serve in the military. As noted earlier, Clinton's backing off conveyed to many that even *he* did not stand strongly behind the policy goal of allowing homosexuals to serve openly. His retreat from his initial position and the new policy that took its place have made the task of implementation both more difficult and less likely to succeed. The new policy and its complications are discussed below.

The New Policy

The new policy, "Don't Ask, Don't Tell, Don't Pursue," was intended to be more permissive than the previous blanket ban on military service. Under the ban, published in 1982, homosexuality was deemed to be incompatible with military service. The new policy makes a distinction between status and conduct, and concedes that sexual orientation alone, absent homosexual conduct, is not grounds for discharge. Homosexual conduct remains a dischargeable offense. The "Don't Ask" feature prohibits some previously permitted behaviors, such as recruiter questions about sexual orientation. Such questions served to screen out at least some homosexuals before entry and provided an easy justification for discharging those homosexuals who lied, entered, and were later discovered. But the ways in which telling has been handled to date under the new policy reinforce the underlying belief of incompatibility that carries over from the 1982 statement. (See chapter 3 for further discussion of this policy.)

From an implementation perspective, the policy changes that have occurred since Clinton proposed to lift the ban have complicated implementation in a number of ways, discussed below.

Implementation Problems

In key ways, the new policy violates implementation tenets to (1) make the policy simple, (2) act quickly, (3) convey the change in terms compatible with military culture, and (4) build in sanctions and enforcement mechanisms to support implementation. Indeed, the policy is characterized by complexity, delay, lack of persuasive efforts, and an absence of sanctions and enforcement mechanisms to support implementation.

A complex policy. The new policy is far from simple. Its three components—Don't Ask, Don't Tell, Don't Pursue—are in some ways inconsistent, and together convey a message of ambiguity concerning the legitimacy of acknowledged homosexuals serving in the military (e.g., Pine, 1995).

Central to the policy's confusion is the distinction that is made, then blurred, between homosexual status and homosexual conduct. For example, DoD Directive 1332.14 (1992) states (p. 4-2) that no basis for discharge exists if the member has said that he or she is a homosexual or bisexual or made some other statement that indicates a propensity or intent to engage in homosexual acts. This distinction blurs quickly because of the rebuttable presumption of conduct given status. Under this presumption, any statement of orientation may trigger questions about behavior. The member bears the burden of proving, by a preponderance of evidence, that he or she is not a person who engages in, attempts to engage in, has a propensity to engage in, or intends to engage in homosexual acts (p. 4-3). If these questions are not answered, they may be used as presumptions of such conduct.

The blurring of the distinction between status and conduct ironically should reduce some implementation problems in the field. Because homosexuals will understand that telling is highly risky, few will do so. Thus the need to "deal with" open homosexuality will be further reduced. At the same time, the "Don't Tell" provision will decrease opportunities to decrease antihomosexual prejudice through equal-status, task-oriented contact, because few homosexuals will reveal their status, given the uncertainty of the result.

Implementation delay. The DoD has not acted quickly to implement the new policy. It chose to delay initiating the policy while directives were written to guide field commanders and to ensure that all services carry out policy in the same manner ("Judge tells Pentagon," 1993). Although such a decision may make implementation easier when it begins, it does imply lack of commitment and allows opposition to grow in the interim.

Lack of persuasive efforts. No efforts have occurred to convey the changes in terms compatible with military culture. This largely reflects a now-confused and complex policy to be implemented in an unwilling organization that won the first round on keeping the ban.

Absence of sanctions and enforcement mechanisms to support implementation. To date, the focus of the new policy has been on the behavior of *homosexuals;* there has been virtually no discussion concerning expected behavior on the part of heterosexual personnel. In particular, there has been virtually no discussion concerning the *bounds* on the discretion of commanding officers to initiate investigations of those who claim homosexual status. Indeed, Korb and Osburn (1995) argue that harassment continues. Yet bounded discretion is particularly important with regard to the implementation of this policy, because the recruiter may no longer serve as gatekeeper to prevent homosexuals from entering the military in the first place, as noted above.

There has also been no discussion of the limits of Defense Criminal Investigative Organization actions, such as what constitutes "credible information" in initiating an investigation. Who determines what is an "appropriate" expenditure of investigative resources? Appropriateness is an undefined criterion for determining whether to investigate under the new policy. Nor has there been discussion of any applicable sanctions for harassment of suspected homosexuals that may result in their revealing homosexual status, as Navy Lieutenant Richard Selland alleges in his case ("Maryland Navy officer," 1994).

As noted earlier, all these lapses reduce the likelihood of successful implementation.

Lack of top leader support. Leadership support is critical to implementation. Here, too, the new policy seems to be failing. There appears to be no real leadership support anywhere. The general perception is that Clinton was bullied into *not* lifting the ban. His backing off confirmed widely held suspicions that he never was *really* behind lifting it in the first place.

The nature of the new policy, which is at best an awkward compromise, is unlikely to attract any strong adherents. Those who supported the lifting of the ban see the new policy as little improvement over the policy in effect before Clinton took office. Those who opposed the lifting of the ban view the new policy as an unnecessary compromise.

Lack of leadership responsibility for its own and subordinates' behavior. There has been virtually no discussion concerning the role of military leadership in monitoring subordinates' behavior, despite the clear possibility that the "Don't Ask" policy may increase harassment as a means of making suspected homosexuals tell. Nor has there been any discussion of training. The only reference in the directive on training (DoD Directive 1332.18, 1987, p. 3) states: "Required knowledge includes an understanding of the conduct necessary to maintain high standards of combat effectiveness and unit cohesion." Such vague statements alone are largely useless in implementing policy.

Absence of a monitoring structure or monitoring mechanisms. This important implementation effort has been virtually ignored, despite considerable evidence of past witch-hunts of suspected homosexuals and no change to date in the rate of discharge of personnel for homosexuality (Korb and Osburn, 1995). Nor has there been any effort to influence what will happen if policy is violated. Yet we know that such perceptions are key in the decisions that individuals make to support or undermine implementation.

Conclusion

Essentially, there has been no effort to implement the new policy. When controversial policies are promulgated in the absence of a strong and focused implementation effort, the net effect may be negative. The policy is likely to arouse strong antagonism and unify restraining forces; this opposition remains unchecked by an implementation effort that attaches high costs both to public resistance and to implementation failures.

A 1996 report issued by the Servicemembers Legal Defense Network, a Washington-based lawyers' group that has been monitoring the "Don't Ask, Don't Tell, Don't Pursue" policy, suggests that the net effect of the new policy is indeed negative. The report concludes that the new policy has done little or nothing to make military life easier for gays. Indeed, the report suggests that things may have gotten worse. These findings support earlier ones reported by Korb and Osburn (1995).

These findings point out the critical role that implementation plays in bringing about change in large organizations. The promulgation of a new policy marks the beginning, not the end, of a long and often difficult process, particularly when the policy meets substantial initial resistance. A careful plan that includes the key aspects of successful implementation is an essential component in bringing about change. In the absence of such a plan and a commitment to its continued monitoring, policy stands little chance of effecting change. And, in the worst case, policy mandates in the absence of strong implementation efforts may exacerbate the very problems that the policy was designed to solve.

Summary

Despite widespread antagonism within the military to a policy that would end discrimination on the basis of sexual orientation, lessons from organization theory, implementation research, and the military's own experiences with African Americans suggest that a new policy might be successfully implemented. Success would depend on understanding the military as a large organization with a unique culture, on a carefully developed and actively monitored implementation plan, and on a sense of the importance of success in implementing the mandate.

The "Don't Ask, Don't Tell, Don't Pursue" policy has not succeeded because an implementation plan has not been developed, and no one has demonstrated a stake in the policy's success. Given substantial resistance to the idea of gays serving in the military, the formulation of this policy in the absence of a strong implementation plan has *increased* opposition and undermined the policy's goals.

The history of this policy underlines the critical role that implementation plays in bringing about change in large organizations. Successful implementation in turn requires a level of commitment to success that never existed here. The antagonism and personal pain that appear to have resulted from the nonimplementation of this policy point to the need to view implementation as an essential component of the policy process. If the will, skill, and capacity to mount a meaningful implementation plan are lacking, then policy development is at best a sham and at worst may be harmful to those that the policy seeks to help.

References

Allaire, Y., & Firsirotu, M. (1985). How to implement radical strategies in large organizations. *Sloan Management Review, 26* (3), 19–34.

Bardach, E. (1980). *Implementation studies and the study of implements.* Paper presented at the annual meeting of the American Political Science Association, Washington, DC.

Berry, S., Hawes-Dawson, J., & Kahan, J. (1993). Relevant military opinion. In National Defense Research Institute, *Sexual orientation and U.S. military personnel policy: Options and assessment* (pp. 209–241). Santa Monica, CA: RAND.

Builder, C. (1989). *The masks of war: American military styles in strategy and analysis.* Baltimore: Johns Hopkins University Press.

Burke, J. (1990). Policy implementation and the responsible exercise of discretion. In D. Palumbo & D. Calista (Eds.), *Implementation and the policy process: Opening up the black box* (pp. 133–148). New York: Greenwood Press.

Dalziel, M., & Schoonover, S. (1988). *Changing ways.* New York: American Management Association.

Department of Defense. (1987). *Military training.* (Directive 1332.18). Washington, DC: U.S. Government Printing Office.

———. (1993). *Enlisted administrative separations.* (Directive 1332.14). Washington, DC: U.S. Government Printing Office.

Eisen, M., & Zellman, G. L. (1992). A health beliefs field experiment: Teen talk. In B. C. Miller, J. J. Card, R. L. Paikoff, & J. L. Peterson (Eds.), *Preventing adolescent pregnancy* (pp. 220–264). Newbury Park, CA: Sage.

Elmore, R. (1978). Organizational models of social program implementation. *Public Policy, 26* (2), 185–228.

———. (1980). Backward mapping: Implementation research and policy decisions. *Political Science Quarterly, 94* (4), 601–616.

Goggin, M. L. (1987). *Policy design and the politics of implementation: The case of child health care in the American states.* Knoxville: University of Tennessee Press.

Gottlieb, N., Lovato, C., Weinstein, R., Green, L., & Eriksen, M. (1992). The implementation of a restrictive worksite smoking policy in a large decentralized organization. *Health Education Quarterly, 19* (1), 77–100.

Janz, N., & Becker, M. (1984). The health belief model: A decade later. *Health Education Quarterly, 11,* 1–47.

Judge tells Pentagon to obey ruling on gays. (1993, October 1). *Washington Times,* p. 3.

Kanter, R. M. (1983). *The change masters: Innovation and entrepreneurship in the American corporation.* New York: Simon & Schuster.

Kilmann, R. (1989). A completely integrated program for organizational change. In A. Mohrman, S. Mohrman, G. Ledford, E. Cummings, & E. Lawler (Eds.), *Large-scale organizational change.* San Francisco: Jossey-Bass.

Korb, L., & Osburn, C. D. (1995, March 19). Asked, told, pursued. *New York Times,* p. E15.

Langbein, L., & Kerwin, C. (1985). Implementation, negotiation, and compliance in environmental and safety regulation. *Journal of Politics, 47* (3), 854–880.

Lawler, E. (1989). Strategic choices for changing organizations. In A. Mohrman, S. Mohrman, G. Ledford, T. Cummings, & E. Lawler (Eds.), *Large-scale organizational change* (pp. 255–271). San Francisco: Jossey-Bass.

Ledford, G., Mohrman, S., Mohrman, A., & Lawler, E. (1989). The phenomenon of large-scale organizational change. In A. Mohrman, S. Mohrman, G. Ledford, T. Cummings, & E. Lawler (Eds.), *Large-scale organizational change.* San Francisco: Jossey-Bass.

Levin, M., & Ferman, B. (1986). The political hand: Policy implementation and youth employment programs. *Journal of Policy Analysis and Management,* 5 (2), 311–325.

Levitt, B., & March, J. (1988). Organizational learning. *Annual Review of Sociology, 14,* 319–340.

Maryland Navy officer faces dismissal for disclosing he's gay. (1994, July 6). *Baltimore Sun,* p. 1.

Mazmanian, D. A., & Sabatier, P. A. (Eds.). (1981). *Effective policy implementation.* Lexington, MA: D.C. Heath.

———. (1983). *Implementation and public policy.* Glenview, IL: Scott, Foresman.

McDonnell, L., & Elmore, R. (1987). *Alternative policy instruments.* (RAND Report JNE-03). Santa Monica, CA: RAND Corporation.

McLaughlin, M. (1987). Learning from experience: Lessons from policy implementation. *Educational Evaluation and Policy Analysis, 9* (2), 171–178.

———. (1990). The RAND change agent study revisited: Macro perspectives and micro realities. *Educational Researcher, 19* (9), 11–16.

Mohrman, A., Mohrman, S., Ledford, G., Cummings, T., & Lawler, E. (Eds.). (1989). *Large-scale organizational change.* San Francisco: Jossey-Bass.

National Defense Research Institute. (1993). *Sexual orientation and U.S. military personnel policy: Options and assessment.* Santa Monica, CA: RAND.

Osburn, C. D., & Benecke, M. M. (1996). *Conduct unbecoming: The second annual report on "Don't Ask, Don't Tell, Don't Pursue" violations.* Washington, D.C.: Servicemembers Legal Defense Network.

O'Toole, L. (1989). Alternative mechanisms for multiorganizational implementation: The case of wastewater management. *Administration and Society, 21* (3), 313–339.

Palumbo, D., & Calista, D. (1990). Opening up the black box: Implementation and the policy process. In D. Palumbo and D. Calista (Eds.), *Implementation and the policy process: Opening up the black box* (pp. 3–17). New York: Greenwood Press.

Peters, T. J. (1978, Autumn). Symbols, patterns, and settings: An optimistic case for getting things done. *Organizational Dynamics.*

Pine, A. (1995, February 6). Few benefit from new military policy on gays. *Los Angeles Times,* p. A1.

Prottas, J. (1984). The impacts of innovation: Technological change in a mass transit authority. *Public Administration Review, 16* (1), 117–135.

Rosenstock, I., Stecher, V., & Becker, M. (1988). Social learning theory and the health belief model. *Health Education Quarterly, 15,* 175–183.

Rousseau, D. M. (1989). Psychological and implied contracts in organizations. *Employee Responsibilities and Rights Journal, 2* (2), 121–139.

Schein, E. (1987). *Organizational culture and leadership.* San Francisco: Jossey-Bass.

Tyler, T., & Lind, E. A. (1992). A relational model of authority in groups. *Advances in Experimental Social Psychology, 25,* 115–191.

Walsh, J. (1991). The social context of technological change: The case of the retail food industry. *Sociological Quarterly, 32* (3), 447–468.

Weatherley, R., & Lipsky, M. (1977). Street-level bureaucrats and institutional innovation: Implementing special education reform. *Harvard Educational Review, 47* (2), 171–197.

Weimer, D., & Vining, A. (1992). *Policy analysis: Concepts and practice.* Englewood Cliffs, NJ: Prentice-Hall.

Wilms, W. (1982). Soft policies for hard problems: Implementing energy conserving building regulations in California. *Public Administration Review, 42* (6), 553–561.

Zellman, G. L., Heilbrunn, J. Z., Schmidt, C., & Builder, C. (1993). Implementing policy change in large organizations. In National Defense Research Institute, *Sexual orientation and U.S. military personnel policy: Options and assessment* (pp. 368–394). Santa Monica, CA: RAND.

Zellman, G. L., Johansen, A. S., & Van Winkle, J. (1994). *Examining the effects of accreditation on military child development center operations and outcomes.* (RAND Report MR-524-OSD). Santa Monica, CA: RAND.

Zetka, J. R., Jr. (1991). Automated technologies, institutional environments, and skilled labor processes: Toward an institutional theory of automation outcomes. *Sociological Quarterly, 32* (4), 557–574.

The President, the Congress, and the
Pentagon: Obstacles to Implementing
the "Don't Ask, Don't Tell" Policy

Lawrence J. Korb

Our founding fathers had a profound distrust of government. Therefore, they designed a system that made it very difficult for the federal government to get things done or to change policies, except in times of crises. Madison and his colleagues at the Constitutional Convention created a central government with limited powers and set up three separate branches with overlapping jurisdictions. Moreover, the founding fathers ensured that the government would be open and accountable to the public by establishing a free and open press.

Throughout the more than two hundred years since the states ratified the Constitution, the branches of government have struggled with each other and the public in trying to provide for the general welfare. Most often the contest for power has been between the executive and legislative branches, with the courts serving as umpire or referee and occasionally stepping in to make policy when the other branches proved incapable of handling a critical issue, such as civil rights.

For most of this century, the president or chief executive has been the dominant force within the political system, and the people have come to hold the president responsible for both the good and bad things that happen to the nation. Thus, presidential candidates tell us that they will end the war in Korea (Eisenhower), get the country moving again (Kennedy), bring peace to Vietnam (Nixon), or curb inflation and get the government off our backs (Reagan). In addition, presidential candidates make a host of specific promises on both large (tax cuts, increases in defense spending) and small issues (the Arctic Wildlife Refuge).

But the ability of the president to dominate the political system was never as great as it seemed. Even Franklin Roosevelt, arguably the most powerful chief executive in this nation's history, could not pack the Supreme Court or establish a National Recovery Administration that could withstand Court scrutiny. Moreover, much of the increased power of the presidency in this century has resulted from the numerous crises that this nation faced, namely two world wars, the depression, and the Cold War (see, e.g., Jones, 1994).

When Bill Clinton campaigned against George Bush in 1992, he, like his predecessors on the campaign trail, made many promises on such major issues as deficit reduction, health care, and welfare reform, and on specific issues such as voter registration via driver's licenses, abortion, and family leave. But with the end of the Cold War, there were no longer any major crises confronting the political system, and regardless of who won the 1992 election, it was going to be difficult for that individual to dominate the political system as had the presidents from Roosevelt to Reagan. This return to normalcy was demonstrated in 1990, when President Bush could not even get a majority in his own party to support the deficit reduction package he had negotiated with the Democratic congressional leadership.

It is against this backdrop that Clinton's campaign promise and subsequent attempt to drop the ban on gays and lesbians in the military must be seen. When Clinton assumed the presidency, he was placed into a constitutional system molded by more than two hundred years of history. He was not only the first baby boomer to occupy the nation's highest office, but also the first president in sixty years to take office in a time free of major crisis.

Clinton's struggle to formulate and implement a policy on gays in the military is also a metaphor for how the political process currently works in the nation's capital. An analysis similar to the one presented here could be done on the president's attempt to enact an economic stimulus package or to bring about health care reform. I believe that analysis of such a contemporary issue must be based as much on intuition as on scientific or historical analyses. In my case, this intuition has been gained from my service in the Pentagon dealing with this as well as similar social issues (e.g., women in combat); testifying on the ban on gays in the military before the Senate Armed Services Committee in March 1993; giving testimony and depositions in several cases of individuals who challenged the ban; and being consulted by individuals in the Clinton White House and Pentagon who were charged with developing and implementing the president's policy. As the public record will show, I am not a disinterested observer. I believe that the ban should be dropped unconditionally. Indeed, in his speech announcing the "Don't Ask, Don't Tell" policy on July 19, 1993, Clinton referred to me and my position on the issue (Clinton, 1993).

The Campaign Promise

During presidential campaigns, candidates make many promises. Some are made primarily for strategic and tactical political reasons, such as to garner the support of a particular constituency or differentiate onself from one's opponent, whereas others are made primarily for reasons of conscience or core beliefs. Bush's statements in the 1988 campaign about "no new taxes" are an example of the former, whereas Paul Tsongas's position on the deficit or Mario Cuomo's stand on the death penalty are examples of the latter.

When their personal beliefs clash with political reality, politicians have two options. They can hold to the principle and suffer the consequences, or they can equivocate. Both options are fraught with danger for the elected official. Equivocation can undermine the official's credibility, whereas steadfastness can force him or her to waste scarce political capital in hopeless battles. Often the choice that a politician makes is a result of the intensity of his or her belief in the position.

During the 1992 campaign, Clinton made it very clear where he stood on the issue of gays and lesbians in the military. In three separate speeches and his campaign document, Clinton said he favored lifting the ban. *Putting People First,* the Clinton and Gore (1992) campaign book, called for an immediate repeal of the ban, and Clinton promised to issue an executive order to end the ban as one of the first acts of his administration.

Was Clinton's promise a matter of conviction or a cynical political ploy? There is no doubt that his position on gays in the military benefitted Clinton politically. Gay and lesbian groups raised nearly $4 million for him, and one in seven who voted for Clinton were homosexual (Drew, 1994). One writer even referred to Clinton as the Abraham Lincoln of the gay rights movement (Burr, 1994).

Moreover, his stand on the issue did not hurt him at all in the campaign. Not much attention was paid to the issue by the Republicans or even the military. In fact, some twenty retired generals and admirals endorsed Clinton, including Admiral William Crowe, a former chairman of the Joint Chiefs of Staff, and Lieutenant General Calvin Waller, the deputy commander of Operation Desert Storm.

There is good evidence that this was a matter of some principle with Clinton. If he stands for anything, it is nondiscrimination or civil rights, the pivot point of politics during his upbringing in Arkansas. Moreover, throughout his tenure as governor of Arkansas, he was in the forefront of obtaining equal opportunity for African Americans and women. Clinton had even tried to pass a gay rights bill through the Arkansas legislature, and said that he would have signed the gay rights bill vetoed in 1991 by California Governor Pete Wilson (Cannon, 1993; Solomon, 1993).

But Clinton was not able to implement his promise. Indeed, one might argue that he capitulated on the issue of gays in the military. Clinton's failure on this issue tells us a great deal about the nature of the American political system and the political skills and core of Bill Clinton.

Implementing the Campaign Promise

Clinton's promise to lift the ban on gays and lesbians may not have attracted much attention during the 1992 campaign, but no issue received more attention during the two-month transition period between the Bush and Clinton administrations. Two days after the election, following a Veterans Day event in Little Rock, the president-elect was asked if he intended to carry out his promise. When Clinton replied that he wanted to, he set off a firestorm of reaction led by Sam Nunn (D-GA), the powerful chairman of the Senate Armed Services Committee, and the Joint Chiefs of Staff, led by their chairman, the charismatic Army general, Colin Powell. As a result, during the transition, many of his political advisers and even Senate Majority Leader George Mitchell (D-ME) urged Clinton to drop the issue (Drew, 1994, pp. 42–48).

But Clinton, who was already being accused of ignoring his campaign commitments on major issues such as the middle-class tax cut, decided to ignore their advice. About one hour after taking the oath on January 20, 1993, at the traditional first presidential luncheon at the Capitol, the president told Representative Gerry Studds (D-MA), an openly gay member of the House, that he would indeed lift the ban. "I am going to do it," he told Studds with deep conviction (Burr, 1994, p. 54).

But Clinton never did sign the executive order to lift the ban. Instead, three years into the Clinton presidency, gays and lesbians in the service are arguably no better off than when Clinton took office. Indeed, they may be worse off because now their place in the military is governed by law, not a DoD directive.

Clinton did not capitulate all at once. Rather, his surrender took place in some five stages lasting throughout 1993.

The first step on the road to surrender took place a mere nine days after Clinton took office. On January 29, Clinton announced a compromise. The president told a news conference that by July 15, 1993, the secretary of defense would, after consulting with the Joint Chiefs of Staff, present him with a draft of an executive order lifting the ban. In the interim, the services would no longer ask applicants about their sexual orientation, and avowed or discovered gays and lesbians would be placed in the standby reserve, a military halfway house, with no pay or benefits. Clinton took this position because Nunn, with the support of the Joint Chiefs of Staff, was threatening to enact the current ban—which was based on Defense Department directives, not legislative sanctions—into law by placing it as a rider on the Family Leave Bill (Drew, 1994, p. 47).

The second stage of the Clinton cave-in occurred when he allowed his own Justice Department to appeal the *Meinhold* decision (*Meinhold v. United States Department of Defense,* 1993). On January 28, 1993, the day before Clinton publicly announced his compromise, U.S. District Judge Terry Hatter had ruled not only that Navy Petty Officer Keith Meinhold, an admitted homosexual, had been wrongfully discharged but that the military's ban on homosexuals was illegal because it lacked even a rational basis, the lowest standard needed to justify a government policy.

Rather than declaring victory and going home, Clinton permitted the department not only to appeal Hatter's ruling, but also to argue in this case and several others over the past two years that discrimination based upon sexual orientation is constitutional; is necessary to maintain good order and morale in the military; is based upon professional military judgment; and that the judiciary should defer to the military in this area. In other words, the policy that was in existence before Clinton came into office was correct and should be upheld in court (Burr, 1994, p. 99).

The next step in the cave-in came on July 19, 1993, when Clinton announced an "honorable compromise" on the issue in a speech at the

National Defense University (Clinton, 1993; Devroy, 1993). This new policy, called "Don't Ask, Don't Tell, Don't Pursue" by its supporters, stipulated that the military would no longer ask a recruit about his or her sexual orientation, and that the military could no longer undertake witch-hunts to root out homosexuals without some credible evidence. But individuals would still be discharged for homosexual conduct, which Clinton described as homosexual actions or statements. Clinton's detractors argued that the policy was in effect "Don't Ask, Don't Tell, Don't Read the Fine Print" (Friedman, 1993, p. 1).

The next stage in his backtracking on the issue came in October 1993, when Clinton signed the bill authorizing defense spending for the 1994 fiscal year. This legislation ignored the crux of Clinton's "honorable compromise," that orientation is not a bar to service. It branded homosexuality an unacceptable risk to military order and stipulated that a future defense secretary could ask recruits and service members about their orientation.

The final stage in the retreat came on December 22, 1993, when Clinton allowed his lame-duck secretary of defense, Les Aspin, to issue the rules to codify the "honorable compromise." These rules place the burden of proving that homosexual status did not mean homosexual conduct upon the accused and permit commanders considerable leeway in deciding how rigorously to enforce the ban on homosexual misconduct (Drew, 1994; Scarborough, 1995).

One indication of Clinton's failure to implement a policy comes from analyzing discharge statistics for 1992 and 1994. In 1992, the year before Clinton took office, and in 1994, the year after the new policy went into effect, the rate of discharges for homosexuality was essentially unchanged (Scarborough, 1994). As Dan Coats (R-IN), a member of the Senate Armed Services Committee and a bitter opponent of Clinton's attempt to change the policy, noted, "Anyone who examines this (the honorable compromise) finds no difference from previous policy." His views were seconded by Representative Robert Dornan (R-CA), now chairman of the House National Security Committee's Subcommittee on Military Personnel. According to Dornan, the new policy was "98 percent of the old policy" (Howlett, 1994).

Obstacles to Implementation

The reasons for the failure to implement the policy are both systemic and personal. They may be placed into eight categories.

First, Clinton had to confront the constitutional division of military powers between the executive and legislative branches. The Constitution makes the president the commander in chief, but it gives Congress the power to raise and support armies, to provide and maintain a Navy, and to make rules for the governance and regulation of the land and naval forces.

Thus, Clinton could issue an executive order eliminating the ban on gays and lesbians, as Harry Truman issued an executive order ending segregation in the armed forces. But Congress could enact legislation banning gays and lesbians from the military, as it had passed laws preventing women from serving on combat ships and flying in combat planes.

Nunn was "Mr. Defense" in Congress. Many legislators from both parties took their cues on defense issues from Nunn. And the senior senator from Georgia believed, unlike the president, that status was conduct and that the military was not ready for an immediate de facto policy of permitting gays and lesbians to serve. Nunn let Clinton know during the campaign and the transition that if Clinton signed an executive order, he would try to enact the existing ban into law, probably as an amendment to the family leave bill, which would be brought up in early 1993 (Drew, 1994).

Second, the new president had to deal with the professional military leadership, the Joint Chiefs of Staff. They are a unique American institution. Although the service chiefs and the chairman are appointed by the president and are nominally subordinate to him and the civilian secretary of defense, they have the legal right to go directly to Congress to voice their complaints on presidential policies affecting national security, because of Congress's constitutional role in that area. Moreover, unlike other presidential appointees, the chiefs do not serve at the pleasure of the president and can be fired only for cause. Rather, they serve terms of four years for the service chiefs, two years for the chairman. Except for Powell, whose term expired in September 1993, all of the Clinton chiefs had been appointed in the second half of Bush's term. Therefore, Clinton had virtually no control over what they could say and do on this issue.

Third, Clinton received little assistance on the issue from the professionals at the Department of Justice. The career attorneys in this department have great reverence for precedents, many of which impinged on the issue of gays in the military. For example, the department had often argued that the courts ought to defer to professional military judgment even on social issues such as women in combat or gays and lesbians in the military. During the 1980s and early 1990s, they had supported the deference to professional military judgment in some two dozen cases challenging the Carter administration's sweeping ban on gays and lesbians (for a list of cases, see Burrelli, 1993). Similarly, the department professionals did not like lower courts to decide constitutional issues. Therefore, when Judge Hatter ruled in the *Meinhold* case that the Carter ban was unconstitutional, those professionals were appalled. Here was a "mere" district judge failing to defer to professional judgment and making constitutional law for the entire nation. Therefore, Hatter's ruling, as well as those of his colleagues, had to be challenged. (These reactions were recounted to me by a department attorney during a deposition.) True, Clinton could have pressured Attorney General Janet Reno to overrule her department, but Reno resisted any White House pressure on her "nonpolitical" organization.

Fourth, Clinton was limited by his weak electoral mandate and his lack of military service. With only 43 percent of the popular vote in 1992, Clinton really did not have much political capital to spend on issues that were not the focus of his campaign. Nearly every Democratic member of Congress had outpolled Clinton in his or her district and thus owed him very little.

Conversely, Clinton's avoidance of military service was an issue in the Democratic primaries and the general election. Therefore, he had to tread very lightly on issues affecting the essence of the armed services. Unlike Presidents Kennedy and Bush, who were war heroes, Clinton could not count on public support in open confrontations with the armed services.

Fifth, although the public generally supported Clinton on this issue, the margins were small and most of that support was not very intense (Howlett, 1993). The opposition, however, felt very strongly about the issue. Congressional mail on the issue was about four hundred to one against lifting the ban (Burr, 1994). Moreover, most of those who voted for Clinton did not want him to focus on this issue early in his administration (Drew, 1994).

Sixth, Clinton had to recognize that the people who would be charged with implementing the policy were the most opposed to it. The Joint Chiefs of Staff strongly opposed Truman's 1948 desegregation order and it took twenty-five years before all the services adopted the spirit of the directive. If Clinton forced this policy on the military, there would be very little likelihood that it would really be implemented. Therefore, the president had to try to get a consensus from the Joint Chiefs of Staff if he really wanted to change the policy on gays and lesbians.

Seventh, Clinton had the misfortune of strong, politically savvy and independent individuals in positions from which they could prevent him from implementing his preferred policy. Colin Powell was no mere Army general. He was a national hero who captivated the public during the Gulf War when he told a transfixed nation that the American military would surround Iraq's army, cut it off, and kill it. Throughout 1993, opinion polls showed Powell beating any Democrat or Republican in a race for the presidency in 1996 (Woodward, 1995). Indeed, Bush had been urged to put him on the ticket in 1992 to save his faltering campaign. As the first African American to become a member of the Joint Chiefs of Staff, Powell was a role model to millions of minorities. Finally, Powell was wise in the ways of Washington. He had served as the assistant to the president for national security affairs in the latter part of the Reagan years and as the military assistant to the secretary of defense, and had held high positions in the Department of Energy and the Office of Management and Budget. It was comparatively easy for Powell to outmaneuver the neophytes in the White House on this issue.

Nunn was in a similar category. He, too, had often been mentioned as a presidential candidate and was fiercely independent. He was instrumental in the congressional rejection of President Jimmy Carter's plan to withdraw American ground troops from Korea, and in forcing Carter to raise defense spending far beyond what the president thought necessary. Nunn not only confronted Clinton on this issue, but even voted against his economic package. Unlike many Democrats, Nunn would not support Clinton just to prevent another failed Democratic presidency. Finally, while Aspin was working on the draft executive order in the first six months of the Clinton presidency, Nunn conducted widely publicized, sensational hearings on the issue, which received nearly as much attention as the O.J. Simpson murder trial. These hearings included as its star witness General Norman Schwarzkopf. The hero of the Persian Gulf War told the senators that if the ban were lifted, American troops would be just like the Iraqi troops who sat in the deserts of Kuwait, forced to execute orders in which they did not believe (Drew, 1994).

Finally, the way in which Clinton conducted himself in the White House weakened his ability to implement his policy. The president consistently framed the issue incorrectly. The issue was never status versus conduct, as he continually stated. The real issue for gays and lesbians in the armed forces was status versus misconduct. Nunn stated that being a homosexual meant that one engaged in homosexual acts (see chapter 3). But as long as the conduct was private and did not violate military rules and regulations—that is, it was not misconduct—it should not have been an issue. Yet because Clinton used the formulation of status versus conduct, an admitted homosexual was considered guilty of misconduct, unless the individual could demonstrate that he or she never engaged in homosexual acts, a very difficult thing to do.

The chronic lack of organization in the Clinton White House also became an obstacle. Clinton's original plan called for him to consult with the Joint Chiefs of Staff during the transition and have the executive order ready by inauguration. Clinton could not find time to consult with the Joint Chiefs of Staff himself. Instead, he had John Holum (now the director of Arms Control and Disarmament Agency) do it. The Joint Chiefs of Staff did not believe Holum was a suitable substitute and asked to see Clinton after the inauguration (Balz, 1993). Moreover, the executive order was not even in draft form by the time Clinton took office. Thus, Clinton could not act as quickly as Carter, who signed the executive order granting amnesty to Vietnam draft evaders and deserters immediately after leaving the receiving stand at his inaugural parade.

As Michael Kelly (1995) pointed out in *The New Yorker*, Clinton also suffered from a peculiarly political sort of learning disability. Once inaugurated, he had a hard time distinguishing between his promise to do something and the reality of having done it. As *New Republic* writer Leon Wieseltier noted (cited in Kelly, 1995), Clinton often seems to think he has acted when he has merely spoken. Thus, when he announced the "honorable compromise," he really believed that he had fulfilled his campaign promise to end discrimination against gays and lesbians in the military.

On the gay and lesbian issue, as in others, the Clinton administration developed a chronic susceptibility to adolescent behavior. According to Kelly, this was manifested by sudden reversals of passion, an incapacity for determined consistency, a talent for lapses in attention and judgment, and an inclination to dissemble in a tight spot, as he did when the Joint Chiefs of Staff and Nunn refused to budge on the issue (Kelly, 1995).

Finally, all these factors combined to deprive Clinton of the honeymoon that new presidents ordinarily receive and use to their advantage in pushing controversial policies. Clinton's plan to lift the ban on gays and lesbians met the same fate as his economic stimulus package and health care reform bill.

Conclusion

Implementing controversial changes is difficult even for a strong chief executive. But for an inexperienced and poorly organized president, with a weak mandate and no credibility in the area, it is almost impossible, given the obstacles built into the governmental system. Whether Clinton should have put up more of a battle is an important political and character question. His cave-in to Nunn and the Joint Chiefs of Staff on this issue gave the impression that he could be easily rolled and that he had no core principles. This perception plagues his presidency.

The president and Congress have spoken and the issue is now in the courts. Indeed, in March 1994, a federal district court in New York ruled the "Don't Ask, Don't Tell" policy unconstitutional because it violated the service member's rights to free speech and equal protection under the law. The court called the policy Orwellian because it equated sexual orientation with misconduct, and Hitlerian because it targeted people not for what they have done, but for their status (Serrano, 1995). Eventually the Supreme Court will have to settle the issue.

References

Balz, D. (1993, January 28). A promise that held inevitable collision. *Washington Post,* p. 6.

Burr, C. (1994). Friendly fire. *California Lawyer.*

Burrelli, D. F. (1993). *Homosexuals and U.S. military policy: CRS report for Congress.* (Report 93-52F). Washington, DC: Congressional Research Service.

Cannon, C. H. (1993, July 4). Crucial details postpone Clinton reply to gay ban. *Baltimore Sun,* p. 1.

Clinton, W. J. (1993, July 19). *Remarks by the President at National Defense University.* Washington, DC: The White House, Office of the Press Secretary.

Clinton, W. J., & Gore, A. (1992). *Putting people first: How we can all change America.* New York: Times Books.

Department of Defense (1982). *Enlisted administrative separations.* (Directive 1332.14). Washington, DC: U.S. Government Printing Office.

Devroy, A. (1993, July 20). President opens military to gays. *Washington Post,* p. 1.

Drew, E. (1994). *On the edge: The Clinton presidency.* New York: Simon & Schuster.

Friedman, T. L. (1993, July 25). Clinton's gay policy: Cave in or milestone? *New York Times,* Section 4, p. 1.

Howlett, D. (1993, July 13). Half support, half oppose compromise on gay troops. *USA Today,* p. 8.

———. (1994, July 28). For homosexuals, 98% of the old policy. *USA Today,* p. 11.

Jones, C. (1994). *The presidency in a separated system.* Washington, DC: The Brookings Institution.

Kelly, M. (1995, April 14). A place called fear. *The New Yorker,* 38–44.

Meinhold v. United States Department of Defense, 34 F.3d 1469 (9th Cir. 1994).

Nunn, S. (1993, January 27). Remarks of Senator Sam Nunn (D-GA). *U.S. Congressional Record–Senate,* p. 755. Washington, DC: U.S. Government Printing Office.

Scarborough, R. (1994, November 18). Don't ask, don't tell has little effect on discharges. *Washington Times,* p. 1.

———. (1995, December 29). Pentagon memo removes winning defense for gays. *Washington Times,* p. 3.

Serrano, R. A. (1995, March 31). U.S. judge rejects military's policy on treatment of gays. *Los Angeles Times,* p. 1.

Shilts, R. (1993). *Conduct unbecoming: Gays and lesbians in the U.S. military.* New York: St. Martin's Press.

Solomon, B. (1993, July 24). In juggling the tough ones, Clinton picks pragmatism over principles. *National Journal,* pp. 184–85.

Woodward, B. (1995, September 24). The Powell predicament. *Washington Post Magazine,* 8–29.

Conclusion

Gregory M. Herek, Jared B. Jobe, and
Ralph M. Carney

.

As noted throughout this volume, the current ban on military service by homosexual personnel rests on the notion that the presence of gay men and lesbians will interfere with the military's ability to accomplish its mission. In a marked break from defenses of earlier versions of the anti-gay policy, the debate since 1992 has explicitly acknowledged that gay people are fully capable of serving competently and with distinction, and that they have done so in the past. Rather than questioning the abilities or patriotism of lesbians and gay men, defenders of the current policy have justified it on the basis of heterosexual personnel's expected reactions to serving with homosexuals. They have assumed that such reactions will be (1) strongly negative; (2) translated into behavior that subverts the military's mission; and (3) beyond the military's control.

The preceding chapters clearly show that the first of these expectations may be justified—at least in the short term—but that the last two are not. Although homosexuality remains stigmatized in U.S. society, as discussed in chapters 9 and 10, antigay sentiment is neither monolithic nor unanimous. Many heterosexuals who morally disapprove of homosexual conduct nevertheless believe that gay people should be protected from discrimination. Moreover, a substantial minority of adults express favorable or supportive attitudes toward gay people, attitudes that appear to be fostered by ongoing personal contact with lesbians and gay men.

Granting that many heterosexuals now in the military would strongly oppose ending the current restrictions on service by gay people, most are unlikely to translate their opinions into behavior that would interfere with the military's ability to accomplish its mission. As documented in chapter 8, the link between general attitudes and specific actions is affected by many intervening personal and situational variables. Chapter 13 shows that the DoD can influence many of these factors through the ways that it frames and implements the policy—for example, by establishing clear norms for behavior and ensuring that enforcement of those norms is supported by leadership. Comparative data from domestic paramilitary organizations and foreign militaries suggest that heterosexuals' negative feelings about working with homosexuals have rarely been acted upon in the institutions that most closely resemble the U.S. military (see chapters 6 and 7). Moreover, even if the presence of a homosexual individual were to affect a military unit, that impact is most likely to be observed in reduced social cohesion, which appears to be largely unrelated to a unit's ability to do its job effectively, unlike task cohesion (see chapter 8).

These assertions are not intended to minimize the importance of heterosexuals' expectations and concerns about an integrated military. Such concerns, however, can be expected to dissipate in the face of day-to-day experiences in the work setting and in barracks and latrines (see chapter 11), provided that officers exercise firm leadership and demonstrate that any new policy will be enforced strictly. As explained in chapter 11, adopting a code of professional conduct regarding sexual behavior that is applied equally and consistently to heterosexual and homosexual personnel, regardless of rank and gender, would go a long way toward ensuring that all personnel could perform their duties effectively.

As with the integration of women and racial minorities (see chapters 4 and 5), problems are likely to occur if the ban is lifted. But as those previous integration experiences have shown, such problems can be capably handled by clear organizational policies and strong leadership. Indeed, we find it instructive to look back and consider the dire predictions that were made about the integration of African Americans into the military. The following quote is taken from the report of a 1942 General Board commissioned to consider the integration of African Americans in the Navy.

Enlistment for general service implies that the individual may be sent anywhere—to any ship or station where he is needed. Men on board ship live in particularly close association; in their messes, one

man sits beside another; their hammocks or bunks are close together; in their common tasks they work side by side; and in particular tasks such as those of a gun's crew, they form a closely knit,
highly coordinated team. *How many white men* would choose, of
their own accord, that their closest associates in sleeping quarters,
at mess, and in a gun's crew should be *of another race?* How many
would accept such conditions, if required to do so, without resentment and just as a matter of course? The General Board believes
that the answer is 'Few, if any,' and further believes that if the issue
were forced, there would be a lowering of contentment, teamwork
and discipline in the service. (Navy General Board, 1942, emphasis
added)

Reasonable observers today recognize that the 1942 rationale for opposing racial integration was groundless, although it may have seemed
like common sense to the members of the board at the time. We note,
however, that if the italicized phrases in the quoted passage are changed
slightly—to *how many personnel* and *homosexual,* respectively—the
quote nicely summarizes many recently expressed concerns about removing restrictions on service by gay and lesbian personnel. The chapters in
the present volume, we believe, provide strong evidence that the current
rationale for banning gay personnel is no more justified than was the
1942 argument against integrating African Americans into the military.

Preparing for a New Policy

The chapters of this book, considered as a whole, refute the notion that
the ban on military service by homosexual personnel is necessary for the
military to continue to fulfill its mission. On the contrary, the weight of
the social science data indicates that the military could successfully eliminate the ban if it chose to do so.

In drawing this conclusion, we do not assume that a change in policy
could be expected to occur without incident. Indeed, the high levels of
partisanship and controversy that have surrounded public debate about
a policy change during the past few years have made an uneventful transition to a new policy even less likely than it might have been before 1993
(see, e.g., chapter 14). But there is no reason to believe that the military
would be unable to deal successfully with any problems that might arise.

Toward that end, we propose that the DoD should begin gathering information now—by conducting empirical research and reviewing previous research findings—to develop the knowledge base necessary for successfully implementing a new policy. Based on the present volume's reviews of social and behavioral research, we recommend four broad areas in which the DoD should collect information.

First, empirical studies should be conducted to document military personnel's current attitudes, beliefs, and perceived norms relevant to a new policy. The primary goal of such study should not be to document that opposition to a new policy exists—we can safely assume that it does—but to assess the reasons for such opposition, to understand the perceived importance of sexual orientation policies relative to other regulations and policies, and to identify demographic groups in which special problems might be encountered during the transition to a new policy. The findings could inform plans for implementing a new policy.

A second area to be addressed concerns the specific content of a new policy. How can it be drafted so that it is clear and simple? How can it best be framed in terms compatible with military culture? What specific target behaviors must be elicited from personnel—heterosexual and gay—if a new policy is to be successful? How, for example, will the DoD distinguish behavior that represents a simple expression of private beliefs or personal identity from discriminatory or harassing conduct? In this area, the DoD can start by reviewing relevant findings from the wealth of empirical studies it has conducted concerning race and gender integration. That body of research will provide a useful starting point that can be supplemented by new studies as appropriate.

A third area for empirical inquiry concerns implementation. What sanctions and enforcement mechanisms, on the one hand, and supporting mechanisms and resources, on the other, will most effectively promote adherence to a new policy? What type of surveillance and monitoring of compliance with a new policy will be most effective at different levels in the chain of command? How can the top DoD leadership most effectively signal its support for a new policy? How can the DoD best train and motivate lower-level leaders to address and solve implementation problems? What will be the roles of chaplains, therapists, health care providers and other guardians of confidential information in implementing the new policy? Here again, the DoD's knowledge base from race and gender integration will be a useful starting point.

A final area for systematic consideration concerns the creation of an organizational climate in which heterosexual personnel's preexisting misconceptions and prejudices toward gay personnel can be overcome. What basic information about gay people do heterosexual personnel need if their stereotypes are to be eliminated? How can the DoD best create opportunities for contact between heterosexuals and gay personnel in a supportive environment in which common goals are emphasized and prejudice is negatively sanctioned?

Obviously, these four areas do not represent an exhaustive list. Many additional questions are suggested by the chapters comprising the present volume. What is important is that the DoD prepare now for a new policy, and that it base its preparations on empirical data as much as possible. Social and behavioral science research has played an important role in helping the military to meet the challenges associated with adapting successfully to dramatic new policies in the past fifty years, including racial and gender integration. If and when the military is faced with the prospect of implementing a new policy concerning sexual orientation, empirical research will offer valuable guidance for meeting that challenge as well.

A View Toward the Future

The debate surrounding the policy on homosexual personnel has consistently focused on predictions of dire consequences if gay men and lesbians were allowed to serve. With the publication of the present volume, that aspect of the debate should be put to rest. There is no rational basis for anticipating that heterosexual and homosexual personnel would be unable to serve competently and professionally, side by side. Arguments to the contrary, we believe, not only ignore the available data, but also question the patriotism, dedication, and competence of all military personnel—heterosexuals and homosexuals alike. Such questioning is, in our view, unjustified and unproductive.

Moreover, preserving the ban creates an undesirable exception to the principles upon which the military's command structure is based. Although personnel are expected to place duty to their country and the good of their unit before their personal opinions and concerns, supporters of the ban assume that heterosexual personnel would be unable or unwilling to do so in the face of serving with a homosexual soldier, sailor, or marine. The result is that the command to serve with a gay man or lesbian is effectively defined as an order that heterosexual military personnel cannot and should not be expected to follow. The wealth of data presented in this volume contradicts the belief that they would be unable to follow such an order. The belief that they should not be expected to follow such an order is an indication of the prejudice against gay men and lesbians that persists in the DoD and in many sectors of society.

We realize that this book's demonstration that the current ban *could* be revoked does not mean that it necessarily *will* be eliminated soon. We hope, however, that the national discourse can shift to a new level. Rather than debating the feasibility of allowing openly gay and lesbian personnel to serve, policymakers should argue about whether changing the policy is the right thing to do for practical, economic, and moral reasons. The appropriate question is no longer whether the military *could* change, but whether it *should* change.

We hope that this question will be answered in the affirmative. Military personnel should be judged on their individual merit and competence, not on aspects of their identity or group membership that are unrelated to their ability. We recognize that others will disagree with us in that regard. Such disagreements, we believe, should be based on differences in values rather than implausible predictions of dire consequences resulting from a policy change.

We foresee that the present policy ultimately will be discarded, although we cannot predict when that will happen. As more heterosexuals come to know gay men and lesbians personally, we expect that societal fears and stereotypes about homosexuality will diminish progressively to the point at which sexual orientation becomes an unremarkable demographic characteristic. Once that day arrives, perhaps future military scholars will look back on the current policy as an odd bit of history that, were it not for the resources it wasted and the lives that it damaged, would be simply a source of curiosity.

In the meantime, we recognize that the debate will continue in the courts, Congress, political campaigns, and countless private conversations among citizens. We sincerely hope that the information and insights in the present volume will inform that debate and enlighten its participants.

Reference

Navy General Board. (1942). *Enlistment of men of colored race in other than messman branch*. (General Board Document No. 421, Serial No. 201). Washington, DC: Department of the Navy.

About the Contributors

Jeffrey E. Barnett, Psy.D., is a licensed psychologist in private practice in Annapolis, MD. He also serves as a clinical assistant professor in the departments of psychiatry and pediatrics at the University of Maryland School of Medicine. He is a former psychologist for the U.S. Army's Special Operations Command. He writes and lectures frequently on ethics issues and dilemmas confronting health care providers.

Ralph M. Carney, Ph.D., is a social psychologist with the Defense Personnel Security Research Center. He is a member of the APA Division of Military Psychology and a Fellow of the Inter-University Seminar on Armed Forces and Society. He was co-editor of *Citizen Espionage* (1994) and author of a chapter in that book on the social history of treason. His current work concentrates on trust, betrayal, and patriotism in the postmodern world.

Paul A. Gade, Ph.D., is chief of the Organization and Personnel Resources Research Unit at the U.S. Army Research Institute for the Behavioral and Social Sciences. He is a Fellow of the American Psychological Association and the Inter-University Seminar on Armed Forces and Society and past president of the APA Division of Military Psychology.

Gregory M. Herek, Ph.D., is a social psychologist at the University of California at Davis. His empirical research has included studies of heterosexuals' attitudes toward gay men and lesbians, violence against lesbians and gay men, and public attitudes concerning the AIDS epidemic. He has published numerous scholarly articles on these topics, and he co-edited *Hate Crimes: Confronting Violence Against Lesbians and Gay Men* (1992). He is also co-editor of the annual, *Contemporary Perspectives on Lesbian and Gay Psychology*. In 1993, he testified on behalf of the American Psychological Association (APA), American Psychiatric Association, and four other national professional associations at hearings on gay people and the military conducted by the House Armed Services Committee. He has also assisted the APA in preparing amicus briefs in court cases challenging the constitutionality of military policies excluding lesbians and gay men, and has served as consultant and expert witness for numerous legal cases involving the civil rights of gay people. A Fellow of the APA and the American Psychological Society (APS), he received the 1992 Outstanding Achievement Award from the APA Committee on Lesbian and Gay Concerns. In 1989, he was the first recipient of the APA Division 44 annual award for distinguished scientific contributions to lesbian and gay psychology.

Peter D. Jacobson, J.D., M.P.H., is senior behavioral scientist at the RAND Corporation, where he focuses on health policy research. He specializes in the relationship between law and health care. His recent work includes a study of the diffusion of new technology in radiology as a function of defensive medicine, a study assessing state-level health care reform initiatives, and a case study of the political evolution of antismoking legislation. He also teaches courses on law and social science at the RAND Graduate School and on law and epidemiology at RAND and the UCLA School of Public Health. In his current work, he is assessing the ways in which judicial decisions influence health care policy, and the ways in which antismoking laws and ordinances are enforced. Before working at RAND, he worked in the U.S. Department of Health and Human Services and at Pittsburgh's Neighborhood Legal Services Association.

Timothy B. Jeffrey, Ph.D., is chairman of the Psychology Department of University Hospital at the University of Nebraska Medical Center. He is a former military psychologist and a past president of the APA Division of Military Psychology.

Jared B. Jobe, Ph.D., is chief of the Adult Psychological Development Cluster at the National Institute on Aging of the National Institutes of Health. A former U.S. Army research psychologist, he has served both on active duty (he is retired from the Army at the rank of captain) and as a civilian, conducting research on health, cognition, personality, and training of military personnel. He is a former president of the APA Division of Military Psychology and is a Fellow of that division as well as the APA Divisions of Experimental Psychology and Health Psychology. He is the author or co-author of more than one hundred and fifty articles, book chapters, technical reports, and conference presentations. He was guest co-editor of a special issue of *Applied Cognitive Psychology* (1991), and was co-editor of *Cognitive Testing Methodology* (1985).

Edgar M. Johnson, Ph.D., is the director of the U.S. Army Research Institute for the Behavioral and Social Sciences and the chief psychologist of the U.S. Army. He is also a Fellow of the American Psychological Association, the American Psychological Society, and the Human Factors and Ergonomics Society.

David E. Kanouse, Ph.D., is a social psychologist whose applied research has often focused on HIV- and AIDS-related issues. He has studied AIDS-related risk behavior in various populations, including adolescents, female prostitutes, and gay and bisexual men. He has also conducted research on health status and treatment outcomes in HIV-infected populations. A senior behavioral scientist at RAND since 1976, he served as head of RAND's Behavioral Science Department from 1989 to 1990.

Michael R. Kauth, Ph.D., is assistant director of the Inpatient Post-Traumatic Stress Disorder Unit and director of mental health services with the Infectious Disease Section at the New Orleans Veterans Affairs Medical Center. He is also clinical assistant professor in the Department of Psychiatry at the Louisiana State University School of Medicine. He is a member of the editorial board of *Assessment* and has authored more than thirty articles, book chapters, and reports.

Paul Koegel, Ph.D., is a behavioral scientist in the Social Policy Department at RAND. An anthropologist by training, he has examined the adaptation of socially and/or economically marginal groups in American society. For the last decade, his main interest has been homelessness. In addition, he has researched issues related to drug and mental health policy, and ways in which the military confronts pressing social policy issues. In the latter regard, he was involved in RAND's examination of homosexuality and the military and is participating in a study of how the careers of minority and female officers compare with those of their white male counterparts.

Lawrence J. Korb, Ph.D., is director of the Center for Public Policy Education and Senior Fellow in the Foreign Policy Studies Program at the Brookings Institution. Before joining Brookings, he served as dean of the Graduate School of Public and International Affairs at the University of Pittsburgh; as vice president of corporate operations at the Raytheon Company; and as assistant secretary of defense for manpower, reserve affairs, installations, and logistics. He has held several academic positions, including associate professor of government at the U.S. Coast Guard Academy and professor of management at the U.S. Naval War College. He served on active duty for four years as a Naval flight officer and retired from the Navy Reserve with the rank of captain.

Dan Landis, Ph.D., is professor of psychology and director of the Center for Applied Research and Evaluation at the University of Mississippi. During 1994–

95, he was visiting professor and, for 1995–96, he was appointed Shirley J. Bach Visiting Professor at the Defense Equal Opportunity Management Institute at Patrick Air Force Base, Cocoa Beach, Florida. He is the founding and continuing editor-in-chief of the *International Journal of Intercultural Relations* and author of more than one hundred articles, book chapters, and technical reports.

Janet Lever, Ph.D., is associate professor of sociology at California State University, Los Angeles. She is also a consultant-in-residence at RAND, where she researches sex and health policy issues. For the past twenty years, Dr. Lever has taught sociology at Yale, Northwestern, the University of California (at Los Angeles and San Diego), USC, and CSULA. Since 1991, she has coauthored (with Pepper Schwartz, Ph.D.) *Glamour* magazine's monthly "Sex and Health" column.

Robert J. MacCoun, Ph.D., is associate professor of public policy at the Graduate School of Public Policy at the University of California at Berkeley. From 1986 to 1993, he was a behavioral scientist at RAND and a faculty member at the RAND Graduate School of Policy Studies. A social psychologist by training, he studies small-group behavior, individual and group decision-making, dispute resolution, formal and informal social control of risky conduct, and citizens' evaluations of government policies.

Theodore R. Sarbin, Ph.D., is the senior editor of *Citizen Espionage: Studies in Trust and Betrayal* (with Carney and Eoyang, 1994) and is involved in efforts to understand and control computer crime. One of the early developers of narrative psychology, he continues to publish articles to support the notion that we live in a story-shaped world. He is a longtime contributor to the scientific study of hypnosis and has participated in the recent controversy about the validity of the repressed memory doctrine. His bibliography lists more than two hundred publications in social psychology, hypnosis, metaphor, imagination, emotional life, narratology, social constructionism, espionage, abnormal psychology, and criminology.

David R. Segal, Ph.D., is professor of sociology and director of the Center for Research on Military Organization at the University of Maryland at College Park and is president of the Inter-University Seminar on Armed Forces and Society. His books include *Peacekeepers and Their Wives* and *Recruiting for Uncle Sam.*

Lois Shawver, Ph.D., is a clinical psychologist who maintains a practice as a psychotherapist and teaches at the California School of Professional Psychology. She publishes on a wide range of psychological issues and is the author of *And the Flag Was Still There: Straight People, Gay People, and Sexuality in the U.S.*

Military (1995). She also serves as an expert in legal cases dealing with the question of bodily modesty, both between men and women and between people of the same gender. Her testimony that heterosexual modesty discomfort would be minimal if the ban on gays were lifted was accepted by the Canadian military, which subsequently dropped the ban. She has submitted similar testimony for many gays and lesbians undergoing discharge proceedings in the U.S. military.

Patricia J. Thomas, M.S., is a supervisory research psychologist at the Navy Personnel Research and Development Center in San Diego. Since 1975, the major focus of her research has been women and minorities in the Navy. She served on the 1990–91 Navy Women's Study Group, was science adviser to the Chief of Naval Personnel (1992), and has been research advisor to the Secretary of the Navy's Standing Committee on Military and Civilian Women since 1992. She is conducting research on sexual harassment in the Navy and the integration of women into combatant ships.

Marie D. Thomas, Ph.D., is a professor of psychology at California State University, San Marcos. Before joining the CSU faculty, she was a personnel research psychologist at the Navy Personnel Research and Development Center and associate professor of psychology at the College of Mount St. Vincent in New York City. Her research interests include gender and ethnicity issues. She is principal coinvestigator on a project on women's friendships and is interested in studying friendships between lesbian and heterosexual women.

Gail L. Zellman, Ph.D., is a social and clinical psychologist at RAND with research interests in a range of youth policies, organizational behavior, and implementation of innovations. She worked on RAND's early studies of implementation of innovations in educational organizations. Her ongoing work on military child care provided her with firsthand experience with military personnel and military culture. In RAND's DoD-funded study of sexual orientation and U.S. military personnel policy, she designed protocols for focus groups of military members, conducted a number of groups, and analyzed the results, in addition to her work on implementation of a new policy. She is leading a study of health care system response to prenatal substance exposure, which builds on her work on child abuse reporting.

Index

Italicized page numbers refer to figures and tables.

personnel policy (*continued*)
132; of laissez-faire, 119–20; of limited tolerance, 118–19; military panel review of, 8; of no-restriction, 117–18; policy versus practice in, 108, 111, 118–21, 125, 148–49; status equals conduct in, 22, 25–27, 33–34, 39; of tolerance, 121–25. *See also* ban on homosexuals; "Don't Ask, Don't Tell, Don't Pursue" (policy); implementation of policy; nondiscrimination policy

Personnel Security Research Center (DoD), 177

Peters, T. J., 277

Pettigrew, T. F., 185

Pexton, P., 187

Pinch, F. C., 109, 116–17

Pine, A., 192, 283

Pittsburgh Men's Study, 26, 31–32

Playboy, readers' survey from, 24

Pliske, R. M., 127

Plummer, K., 210

police departments: acknowledged homosexuals in, *137, 165*; antihomosexual feelings in, 139–40, 143; attitudinal changes in, 144–46; characteristics of homosexuals in, 141–42; heterosexual concerns in, 143–44, 150; homosexuals' behavior in, 140–41; and impact of policy change, 137–39; information gathered on, *134, 135, 136*; military compared to, 132–34, 149–51. *See also* nondiscrimination policy

policy: complexity of, 282–83; contents of, 306; context for, 266–69; design for, 269–71, 274–77; framing of, 298–99; instruments for, 270–71; preparation for, 305–7; sanctions for noncompliance with, 276–77; shift in, 281–82. *See also* implementation of policy; mandates; personnel policy

politics: and campaign promises, 290–300; metaphor for, 291; military's connections with, 11

Pomeroy, W. B., 16, 18–19, 229

Ponse, B., 201–2

Pope, K. S., 254–55, 257

Porteous, S., 236

Portugal, military policies in, 119

positive contact experiences, and attitudinal change, 145

Powell, Gen. C.: on exclusion of homosexuals, 3, 94–95, 239n. 1; on privacy argument, 94–95, 226–27, 239n. 1; role of, 293, 298; on support for Clinton, 268

pregnancy: and assignment restrictions, 82; discharge for, 72; and homosexuality investigation, 73

prejudices: basis of, 93; versus conduct, 168–71, *170*; court cases on, 48, 51–52; versus external threats to group, 163; persistence of, 100; response to, 48; ways to counter, 185–86, 213–17, 306–7. *See also* attitudes; segregation; stereotypes

Presidential Commission (Bush), 68–69, 78–79

prisons, sexual behavior in, 25, 34n. 3

privacy argument, 94–95, 226–28, 239n. 1

privacy rights: versus coming out, 108, 123–24; concept of, 248; of heterosexuals, 4, 48–49, 95, 97; in military, 49, 77, 257–58; support for, 190, 226; zone for, 40. *See also* confidentiality; modesty

privilege, and confidentiality, 250, 259–60

professionalism, effects of, 143

Project Clear, 88

propinquity, 161

protected class: criteria for, 53, 58n. 8; homosexuals as, 41–43, 46, 49, 51, 57–58n. 2; resentment of, 144, 147,